Practical Particle Physics

David Michalets

Self-published on **December 26, 2020**

Table of Contents

1 Introduction 4
2 Fundamental Particles 12
3 History of Atomic Model 22
4 Planck's Constant 38
5 Photoelectric Effect 42
6 Particle Pair Production 45
7 Initial Changes to Atomic Model . . 49
8 Einstein and Maxwell 92
9 Gravity 98
10 Mechanism for Force of Gravity . 104
11 Light 126
12 Atomic Equilibrium 152
13 Mass Defect Introduction 172
14 Isotope Data File 176
15 Periodic Table 182
16 Chart Isotope Changes 302
17 Elements Conclusion 304
18 Final Conclusion 310
19 References 312

Introduction

Practical Particle Physics takes a practical approach to the atomic model and its subatomic particles, by beginning with the observational evidence. This process reviews the theoretical aspects of subatomic particle physics, such as quasi-particles, like photon and graviton. These probably arose with the introduction of Einstein's theory of relativity into physics.

Relativity's unique context of a special moving observer's reference frame is reviewed as an application in the narrower scale of particle physics.

The science of particle physics has failed to address a fundamental problem which is commonly called atomic mass defect.

The results of analyzing isotopes across the entire periodic table of 118 elements are presented. The mass defect should be explainable, but it requires a review of the current atomic model.

This book reviews Maxwell's work with electromagnetism and light, and it includes the author's proposed mechanism for Newton's force of gravity.

Light and the fundamental forces of electric, magnetic, and gravitational had been explained using classical physics and without the later introduction of quasi-particles, like a photon or graviton.

The electric and gravitational forces are instantaneous and are difficult to explain as a particle behavior, so the fundamental forces are reviewed.

The origin of anti-particles is also part of this review of the Standard Model.
The author wrote 4 books about physics, before this one.

This book will sometimes explicitly adapt text from the other books, but those books are not required to read this one.

The reader can reference any of those books, if this presentation is not convincing enough without all the supporting data contained in the other books.

An emphasis of this book is the mass defect behavior. Though it was mentioned in a previous book, this book is based on the analysis of the changes in mass of isotopes, covering as many as practical. That exercise started with this book, so other books offer nothing on this topic.

Each book of the 5 has its own focus. Physics covers many phenomena, so there is some cross-over between books in the set.

This is a summary of the previous 4 books. All 5 were self-published using Kindle Direct Publishing so all are distributed by Amazon.

1) Observing Our Universe.

Problems arise when ignoring the context of the observer.

The problem with red shift begins by ignoring we view the universe only from on or near the Earth.

The problem worsens when expecting only a simple Doppler effect can cause a red shift. There are actually 4 distinct mechanisms causing a shift in lines within a spectrum, 2 separate mechanisms for galaxies and 2 for quasars.
Halton Arp observed a discrepancy when comparing red shift values but he failed to explain their mechanisms.

There are many wrong conclusions when measurements affected by 4 mechanisms are assumed to be from just a simple Doppler effect.
The correct interpretation of a red shift is explained.
The new method for calculating a distance to a distant galaxy is explained.

There is no universe expansion, no dark energy, and there was no big bang.

Nearly all galaxy and quasar values based on the red shift mistake are wrong.

The correct use of a distant galaxy red shift for its distance calculation is proposed.

Einstein proposed a special observer for his theory of relativity.

Objects in the universe do not possess this special observer so relativity does not apply to anything except when accepting one is the special observer and so observations are confined within one's reference frame, which is called space-time.

Relativity should be abandoned, as it was an unfortunate diversion from valid physics. For example, space-time enabled a black hole which violates physics., leading to other impossible things like an accretion disk emitting X-rays.

There is no curvature and the universe does not possess a topology.
Mass does not have a velocity limit at c and gravity does not wave.

LIGO tries to observe gravitational waves using equipment on Earth's surface. LIGO ignores their context when claiming to observe the universe for these theoretical waves having no definition in terms of physics making them impossible to detect using a device designed for the wave's physical characteristics. Having none, LIGO tries to detect them indirectly with the expectation whatever they can detect could be what they seek. LIGO is bad science, never having evidence for their claims.

In early 2019, this author concluded LIGO is ignoring a terrestrial source which is triggering their detections from a claimed astrophysical source. This terrestrial source is predictable.

In November 2019, this author gave LIGO explicit predictions of LIGO detections which were confirmed by LIGO reports.

Gravitational waves do not exist.

LIGO has never detected a gravitational wave. Its claims of mergers have never been verified. Every LIGO claim of detection or confirmation must be ignored, including relativity and black holes.

The first book references studies published by others which explain that dark matter is being proposed wherever a magnetic field is ignored. A magnetic field is not visible but it can be measured when it affects light passing through it.

Numerous studies have confirmed there is no need for dark matter. Perhaps it persists because the Lambda – Cold Dark Matter cosmological model requires dark matter for its failed attempt to explain the evolution of the universe after its beginning at the big bang. Numerous people, including this author in several books, have explained why there was no big bang.

CERN and LHC are on a fool's errand looking for a dark matter particle which does not exist.

2) Cosmology Transition

The first book noted errors in astronomical data. This second book describes how to fix the data and offers a framework for the organization of data for galaxies and quasars. There is no such framework now.

The transition includes a new solar model from Dr. Robitaille based on the observational evidence, so its mechanism uses thermal radiation from condensed matter, not fusion in an impossible internal equilibrium deep within a gaseous sphere, a model for the spiral galaxy rotation from Dr. Scott, and the author's model for a quasar.

3) Cosmology Connections

The neglected electrical connections in the universe are described, from within our solar system, to beyond. Often light is considered to be from a heat source, like a light bulb having a filament.

Actually, there are many sources of measurable sources of synchrotron radiation spanning from high energy gamma rays to low energy infrared or radio. Synchrotron radiation results from an electric current being diverted by a magnetic field so a broad spectrum of electromagnetic radiation results.

The third book identifies a) mistakes with thermal for synchrotron radiation, and b) sources of magnetic fields and synchrotron radiation.

For example, an X-ray source can be mistaken as an impossibly hot object when it really is synchrotron radiation.

A fictitious black hole with an impossible hot accretion disk is the most blatant example of this mistake.

There are many X-ray point sources in the universe and every one of them is generating synchrotron radiation by an electrical mechanism, not heat.

4) Redefining Gravity

Isaac Newton defined the force of gravity but not the force.

This book defines the mechanism. Some of this material was adapted for this subsequent book.

James Clerk Maxwell lived long after Newton so Newton could not apply Maxwell's work done in the future. Physicists made the mistake of assuming relativity's space-time redefined gravity as curvature. This mistake must be fixed.

The fourth book describes the substantial evidence for Newton's force including Kepler's laws of motion which are based on ellipses around a center of gravity.

The book proposes a change to Kepler's third law so it applies to moons and exoplanets. As stated now, the third law is limited to only objects in orbit around the Sun.

Relativity and its space-time should be removed from physics.

2 Fundamental Particles

In an updated atomic model for particle physics, only a permanent particle having a measurable mass is a fundamental particle.

The Standard Model for particle physics has different criteria. A comparison between atomic models is in the section about their history.

This section sets the context for this book.

This book proposes the 2 fundamental subatomic particles are the proton and electron. The rare neutrino and muon are also described.

A small number of subatomic particles are observed in nature, meaning outside of particle accelerators.

These are the electron, proton, neutron, and neutrino. The muon barely qualifies because it exists in nature only as a result of atmospheric particles colliding with cosmic rays, which are the results of distant astrophysical particle accelerators.

Each natural subatomic particle's current basic description is provided. More details for each particle will follow later.

2.1 Definition of several terms in this section

Elementary or fundamental particle, from Wikipedia:

In particle physics, an elementary particle or fundamental particle is a subatomic particle with no substructure, i.e. it is not composed of other particles.

Observation:

The simple rule for an elementary particle is: it exhibits no other particles.

2.2 Electron

Excerpt from Wikipedia:

The electron is a subatomic particle, symbol e− or β−, whose electric charge is negative one elementary charge. Electrons belong to the first generation of the lepton particle family, and are generally thought to be elementary particles because they have no known components or substructure.

(Excerpt end)

Observation:

The electron is elementary only because particle colliders have been unable to find a way to break one.

Perhaps someday, someone will accomplish this feat, resulting in an electron getting a new classification, depending on the fragments.

Even the use of "generally thought" confirms this rule is based on a judgment.
Particle physics needs a rule which is based on some physical evidence rather than on only the limits of accelerator technology.

2.3 Quark

From Wikipedia:

"A quark is a type of elementary particle and a fundamental constituent of matter. Quarks combine to form composite particles called hadrons, the most stable of which are protons and neutrons, the components of atomic nuclei. Due to a phenomenon known as color confinement, quarks are never found in isolation."

Observation:

Despite this description, a neutron is not stable. A neutron outside a nucleus disintegrates into its 2 parts, proton and electron, in a few minutes. When a nucleus has too many neutrons, one of them will eject its electron leaving only the proton. This is known as beta minus decay.
A quark is not consistent with the electron and proton particles because it is "never found in isolation."

A quark is only a fragment remaining after the destruction of a proton. Otherwise, it is "never found."

There is no justification for it to be fundamental, rather than worthless debris.

Quarks contribute nothing to our understanding of an atom.

This book is titled Practical Particle Physics because one of its goals is explaining atomic mass defect to advance our understanding of an atom's behaviors. Quarks have no role in any known atomic behavior.

If someone writes about particles observed by users of the LHC, quarks should be in that book. After this preliminary section, quarks do not belong in this book.

2.4 Proton

Excerpt from Wikipedia:

A proton is a subatomic particle, symbol p or p^+, with a positive electric charge of +1e elementary charge and a mass slightly less than that of a neutron."

Although protons were originally considered fundamental or elementary particles, in the modern Standard Model of particle physics, protons are classified as hadrons, like neutrons, the other nucleon.

Protons are composite particles composed of three valence quarks: two up quarks of charge +2/3e and one down quark of charge −1/3e. The rest masses of quarks contribute only about 1% of a proton's mass. The remainder of a proton's mass is due to quantum chromodynamics binding energy, which includes the kinetic energy of the quarks and the energy of the gluon fields that bind the quarks together. Because protons are not fundamental particles, they possess a measurable size; the root mean square charge radius of a proton is about 0.84 to 0.87 fm (or 0.84×10^{-15} to 0.87×10^{-15} m).

In 2019, two different studies, using different techniques, have found the radius of the proton to be 0.833 fm, with an uncertainty of ±0.010 fm.
***

(Excerpt end)

Observation:
The quarks do not sum up to the mass of a proton. They have only 1%. If quarks are the claimed components of a proton then the claim immediately fails. A proton needs much more than these fragments to be capable of its observed behaviors.
This book treats a proton as a fundamental particle. A proton exhibits no substructure.

A proton can be broken into 3 fragments but they cannot be combined again to get a functional proton. Fragments from a high velocity collision are not components or substructure.

A proton exhibits mass and charge so if the set of fragments is claimed to be its components or substructure then there must be evidence these fragments either share or contribute to the known behaviors of the original particle. For a proton, there is no such evidence for the fragments being legitimate components. The 1 % is not evidence, but instead shows this is a mistake. They are no different than fine particle dust remaining after an explosion. The original object's structure is not found in the dust.

The 3 fragments have no evidence any of them was a component of an important behavior, like mass or charge.

Because they provide nothing to improve our understanding of a proton, quarks should be ignored when explaining behaviors of atoms and its 3 components, electron, proton, and neutron. Quarks in a proton will be considered again, in the section 17 Elements Conclusion.

2.5 Neutron

From Wikipedia:

"The neutron is a subatomic particle, symbol n or n^0, which has a neutral (not positive or negative) charge and a mass slightly greater than that of a proton."

Observation:

The following statement is not in Wikipedia because it is this book's conclusion.

A neutron is a proton having an adjacent electron.

Therefore, a neutron is correctly not a fundamental particle because it is a combination of 2.

2.6 Neutrino

From Wikipedia:

A neutrino (denoted by the Greek letter ν) is a fermion (an elementary particle with spin of 1/2) that interacts only via the weak subatomic force and gravity. The neutrino is so named because it is electrically neutral and because its rest mass is so small (-ino) that it was long thought to be zero.
The mass of the neutrino is much smaller than that of the other known elementary particles.

neutrinos in one of three leptonic flavors: electron neutrinos ($ν_e$), muon neutrinos (ν μ), or tau neutrinos (ν τ), in association with the corresponding charged lepton Although neutrinos were long believed to be massless, it is now known that there are three discrete neutrino masses with different tiny values, but they do not correspond uniquely to the three flavors.

(Excerpt end)

Observation:

A neutrino is elementary only because it was never broken into pieces. There is no observed decay.

The mass remains "tiny" for each neutrino.

This lack of definition is awkward for its "elementary" status.

The neutrino will be described in more detail later, after other atomic behaviors are explained. A neutrino is part of section 12 Atomic Equilibrium.

2.7 Muon

From Wikipedia:

The muon from the Greek letter mu (µ) used to represent it) is an elementary particle similar to the electron, with an electric charge of −1 e and a spin of 1/2, but with a much greater mass. It is classified as a lepton. As with other leptons, the muon is not known to have any sub-structure – that is, it is not thought to be composed of any simpler particles.
The muon is an unstable subatomic particle with a mean lifetime of 2.2 µs.

Muons have a mass about 207 times that of the electron.

The dominant muon decay mode (sometimes called the Michel decay after Louis Michel) is the simplest possible: the muon decays to an electron, an electron antineutrino, and a muon neutrino.

(Excerpt end)

The muon is not consistent with other elementary particles:

a) It decays.

b) Because the muon decays, then those fragments should be considered the muon's components.

c) Despite its decay, the muon remains an elementary particle in the Standard Model. It is even explicitly "similar to the electron" which never decays.

d) Its mass is 207x an electron but it decays into only 1 electron plus 1 muon neutrino plus 1 electron antineutrino.

All types of neutrinos currently have a mass of "small but non-aero."
Therefore, the decay of a muon results in the loss of mass equivalent to more than 206 electrons. This result should be a crisis for particle physics, but I never knew about it until writing this page. The muon is not an important particle.

This book solves the problem of mass defect for particle physics. Solving the muon decay's loss of mass must begin with checking its mass measurement.

As a super-heavy electron, one can wonder what happens during a proton encounter. They could attach to become a heavy neutron or it could orbit like Hydrogen, except gravity also has a role with the different mass ratio. One could wonder why these tests are not mentioned, but with a brief life such tests are unlikely.

The muon is lacking a credible definition and has a very brief life. Its status as an elementary particle cannot be justified, as if it were fundamental to this science.

The muon has no role in any atomic behaviors.

Like quarks, muons should be relegated to whatever method is used to describe anomalies, like debris from particle accelerators, which include muons and quarks.

This book's goal is a practical particle physics with can better explain the behaviors of atoms and the particles within them. Some particles, as just fragments of high velocity collisions and having very brief lives, are out of scope.

3 History of Atomic Model

3.1 Initial models

Here is a brief version using an excerpt from the site AzCemistry:

The first atomic theorist was Democritus, a Greek scientist and philosopher who lived in the fifth century BC. At that time, Democritus found that if a stone was divided in half, the two halves would have essentially the same properties as the whole. And after that, he tried to cut the stone continually into smaller and smaller pieces up to some point where there would be a piece that would be so small as to be indivisible. He called these small pieces of matter "atomos", in Greek it means indivisible.

John Dalton developed the atomic theory around the 1800s. He developed the atomic theory because he disagreed with the theory of atoms that Aristotle had previously proposed. He passed through several experiments and discovered several atomic weights and created symbols for atoms and molecules.

In 1897, a scientist named J.J. Thomson did research to refine Dalton's atomic theory. Joseph John Thomson was a professor of experimental physics. He was successful for developing the atomic theory and received a Nobel Prize in physics in 1906 for his discovery of atomic theory. At that time, Thomson experimented using a cathode ray tube or also called an electron gun. Thomson used a cathode ray tube with a magnet and find that the resulting green beam is made of negatively charged material.

In addition, he also did a lot of research and found that the mass of these particles is almost 2000 times lighter than the hydrogen atom.

From this study, Thomson suggested that Dalton's theory of atoms which said that atom could not be divided into smaller parts was wrong. After that, Thomson conducted a follow-up study and determined that the negative charge of the electron requires a positive charge that can balance both. Thus, he concludes that this negative charge is surrounded by a positively charged material.

A few years later, more precisely in 1911, Ernest Rutherford, one of Thomson's disciples, did some further research on Thomson's plum pudding model. The study was conducted by firing beam from positively charged particles called alpha particles against a very thin layer of gold foil. Since alpha particles had a lot of mass, Rutherford thought that all alpha particles would penetrate directly to the gold foil. At that time Rutherford argued that the alpha particle would penetrate the positively charged material and then would hit the screen detector on the other side.
But that hypothesis did not match with what Rutherford had predicted. Some of the alpha particles penetrated the gold layer, but some of them were deflected by the gold foil and then hit a detector at another location. Some of them even returned straight back to the path they took.

After this research, Rutherford argued that this alpha particle must hit something that was very small, dense, and positively charged so that there were some alpha particles that went straight back. From this experiment, Rutherford also concluded that atoms were composed mostly of empty spaces and the existing positive charge is not evenly distributed within the atom but squished into a tiny nucleus in the center of the atom.

Although Rutherford's theory can show that atoms have a nucleus that is positively charged and surrounded by a negative electron, this theory also has a weakness. The weakness is that Rutherford's theory cannot explain why electrons do not fall into the nucleus.

In 1913, the Danish physicist and also a student of Ruhterford, Neils Bohr repaired Rutherford's atomic theory through his experiments on the spectrum of hydrogen atoms. This experiment managed to give a picture that electrons are occupying the area around atomic nucleus. Bohr's explanation of a hydrogen atom involves a combination of Rutherford's classical theory and the quantum theory of Planck, expressed by four postulates, as follows:

There is only a certain set of orbits that is allowed for one electron in a hydrogen atom. This orbit is known as a stationary motion (settling) electron and is a circular path around the nucleus. The path, which is also called the atomic shell, is a circular orbit with certain radius. Each path is marked by an integer called the principal quantum number (n), starting from 1, 2, 3, 4, 5, and so on and denoted by the symbols K, L, M, N, O, and so on. The first path with n = 1 is named shell K. The second path with n = 2 is named shell L, and so on. The larger the n value it means that is farther from the nucleus hence the greater the electron's energy orbiting the skin.

As long as the electron is in the stationary path, the energy of the electron is still remaining so there is no energy in the form of radiation that is emitted or absorbed. Electrons can only move from one stationary path to another stationary path. In this transition, a certain amount of energy is involved, the magnitude corresponding to the Planck equation, $E_2 - E_1$ (ΔE) = hf

The allowed stationary path has a magnitude with certain properties; particularly the property is called angular momentum. The magnitude of the angular momentum is a multiple of h / 2p or nh / 2p, where n is an integer and h is the Planck constant.

According to the Bohr's atomic model theory, electrons surround the nucleus at certain paths called electron shells or energy levels. The lowest energy level is the deepest electron shell; the outer layer has the bigger shell number and higher energy level.

The summary of Bohr's atomic model theory is that atom consists of atomic shells as a place for electron to move. Despite this new invention, this theory also had a weakness that is this atomic theory could not able to explain the colors spectrum of atom that consisted of many electrons.

[Modern Atomic Model]

In 1927, Erwin Schrodinger, an Austrian scientist, put forward an atomic theory called quantum mechanical model of the atom. This theory uses mathematical equations to explain the possibility of finding electrons in certain positions. The quantum mechanical model has similarities with the Niels Bohr atomic theory in terms of energy levels or atomic skins, but differs in terms of their shape or orbit.
In the atomic theory of quantum mechanical model, the position of electrons is uncertain. The thing that can be determined is about to predict the odds of the location of the electron. This model can be portrayed as a nucleus surrounded by an electron cloud. Where in the most dense clouds, there is the greatest possibility for the discovery of electrons, and vice versa, the least electrons are found in less dense cloud regions. Thus, this theory model also explains the concept of sub-energy level.

The formulation of Erwin Schrodinger was very difficult to understand by scientists at that time. Schrodinger's theory is similar to the solar system whose orbits are erratic and the sun is at its core. For the discovery of the atomic theory of quantum mechanics, Erwin Schrodinger received physics Nobel Prize in 1933.

After understanding the atomic development theory, we also have to understand more about the subatomic substances invention. The invention of subatomic substances was a great discovery to help scientists developed the atomic model theory. Hence, we will explain more about the history of subatomic substances below.

(Excerpt end)

Observation:

I would not characterize the solar system orbits, being ellipses around a center of gravity, as being erratic.

Gravity maintains the solar system's stability. Gravity is a relatively weak force for the relatively slow objects, like planets and asteroids in orbit at a significant distance. In the atom, gravity is assumed negligible, so the electrostatic force is assumed to maintain stability within the atom which has a smaller volume for its much smaller particles. The positive nucleus is compact compared to the orbits of negative electrons.

The solar system is a misleading comparison to an atom.

An atom can absorb energy, in the form of electromagnetic radiation, which is held in its electrons by their velocity in orbit. An atom can emit
Energy when its electrons drop to a lower energy state.

A solar system has no behaviors for holding and releasing energy from its collection of bodies.

(Excerpt continued:

[Subatomic Particles]

The subatomic substances are the neutrons and protons.

In science nowadays, proton determines the atomic number. Meanwhile, atomic nucleus which has a very large weight compared to electron will determine the mass of an atom.
The weight of neutrons and protons usually has the same amount. Protons have a positive charge while neutrons are uncharged (neutral). In the history of chemistry we know that atoms have substances, especially parts of protons and neutrons that are interlocking with each other. At the beginning of the discovery of protons and neutrons were obtained by various experiments by scientists.

In 1886, a German physicist named Eugen Goldstein invented the anode or proton rays. At that time, the essence of the cathode ray had not determined yet, but Goldstein conducted an experiment with a cathode ray tube and found the following facts.
If the cathode is not hollow, it turns out that gas behind the cathode remains dark. However, when the cathode is given a gas hole behind, the cathode will emit light.
This indicates the presence of radiation coming from the anode, which passes through the hole at the cathode and permits the gas behind the cathode.

The radiation is called an Anode ray or positive ray or canal ray. The experimental results show that the canal rays are radiation particles that are positively charged. The radio-graphic particles turn out to depend on the type of gas in the tube. It means, if the gas in the tube is replaced, it will produce particles of light canal with different size. The smallest irradiation particles are obtained from hydrogen gas. These particles are then called protons.

Mass of 1 proton = $1.6726486 \times 10^{-24}$ grams = 1 sma

The charge of 1 proton = $+1 = +1.6 \times 10^{-19}$ C

The charge and mass of the light particles composed of other gases and is always in a spherical multiplier of the mass and the charge of the protons, so it is suspected that the particles consist of protons. Then in 1919, Rutherford discovered that protons formed when alpha particles were fired at the nucleus of a nitrogen atom. The same thing happened to the firing of another atomic nucleus. This proves that the atomic nucleus consists of protons as alleged by Goldstein.

Neutron was invented by James Chadwick along with Ernest Rutherford in 1932 using alpha rays, but that is still only an assumption, and its existence has been suspected by Aston since 1919. In that year, Aston invented the mass spectrometer, which is the instrument that can be used to determine mass Atoms or molecules.

With the instrument, Aston discovered that atoms of the same element can have different masses. This phenomenon is called isotope. It was also found that the mass of an atom was not the same as the number of protons. Many atom mass are about twice of the proton mass. Based on these two facts, Aston suspected that the existence of neutral particles in atoms with different amount even though the elements are the same.

Later in 1930, W. Bothe and H. Becker fired at beryllium nuclei with alpha particles and discovered a high-penetrating particle radiation.
In 1932, James Chadwick proved that the radiation consists of neutral particles whose mass is almost equal to the mass of protons. Because they are neutral, they are called neutrons. Further experiments prove that neutrons are also fundamental particles of the atomic constituents.

Mass 1 neutron = $1.6749544 \times 10^{-24}$ gram = 1 sma

Neutrons are uncharged (neutral)

(Excerpt end)

Observation:
Through these many experiments, the basic subatomic particles were identified and mass values were assigned in grams.

The mass values for both proton and neutron had 8 digits after the decimal.

The mass value for the electron was only estimated at 2000 to 1 proton.

Physicists tried to improve the precision of these values.

For reference, here is the description, from Wikipedia, of mass spectroscopy:

Mass spectrometry (MS) is an analytical technique that measures the mass-to-charge ratio of ions. The results are typically presented as a mass spectrum, a plot of intensity as a function of the mass-to-charge ratio.

Mass spectrometry is used in many different fields and is applied to pure samples as well as complex mixtures.
A mass spectrum is a plot of the ion signal as a function of the mass-to-charge ratio. These spectra are used to determine the elemental or isotopic signature of a sample, the masses of particles and of molecules, and to elucidate the chemical identity or structure of molecules and other chemical compounds.
In a typical MS procedure, a sample, which may be solid, liquid, or gaseous, is ionized, for example by bombarding it with electrons. This may cause some of the sample's molecules to break into charged fragments or simply become charged without fragmenting.

These ions are then separated according to their mass-to-charge ratio, for example by accelerating them and subjecting them to an electric or magnetic field: ions of the same mass-to-charge ratio will undergo the same amount of deflection. The ions are detected by a mechanism capable of detecting charged particles, such as an electron multiplier. Results are displayed as spectra of the signal intensity of detected ions as a function of the mass-to-charge ratio. The atoms or molecules in the sample can be identified by correlating known masses (e.g. an entire molecule) to the identified masses or through a characteristic fragmentation pattern.

(Excerpt end)

Excerpt continued for Mass-to-Charge Ratio:

The mass-to-charge ratio (m/Q) is a physical quantity that is most widely used in the electrodynamics of charged particles, e.g. in electron optics and ion optics.

It appears in the scientific fields of electron microscopy, cathode ray tubes, accelerator physics, nuclear physics, Auger electron spectroscopy, cosmology and mass spectrometry. The importance of the mass-to-charge ratio, according to classical electrodynamics, is that two particles with the same mass-to-charge ratio move in the same path in a vacuum, when subjected to the same electric and magnetic fields. Its SI units are kg/C. In rare occasions the thomson has been used as its unit in the field of mass spectrometry.

Some disciplines use the charge-to-mass ratio (Q/m) instead, which is the multiplicative inverse of the mass-to-charge ratio. The CODATA recommended value for an electron is Q/m = $-1.75882001076(53) \times 10^{11}$ C/kg

(Excerpt end)

Observation:

This method tries to measure the mass of a charged particle in kg.

Eventually, another unit was introduced, the atomic mass unit. This unit is described in a subsequent section.

3.2 Standard Model

Matter is every object we can see or measure. Each object is radiating, reflecting, or absorbing energy. Each can also be moving when it carroes kinetic energy.

Any particle of matter holding an electrical charge is also called plasma. Plasma has unique behaviors compared to matter having no charge, because of the interaction of electric and magnetic fields.

Matter consists of atoms and their molecules, as well as any subatomic particles having mass, such as electron, proton, and neutrino.

Excerpt from Wikipedia:

In classical physics and general chemistry, matter is any substance that has mass and takes up space by having volume. All everyday objects that can be touched are ultimately composed of atoms, which are made up of interacting subatomic particles, and in everyday as well as scientific usage, "matter" generally includes atoms and anything made up of them, and any particles (or combination of particles) that act as if they have both rest mass and volume.

(Excerpt end)

There has been a standard model for an atom for a long time. The Large Hadron Collider, LHC, has been used frequently to learn about the subatomic particles declared to be part of the Standard Model.

Excerpt from Wikipedia:

The current state of the classification of all elementary particles is explained by the Standard Model, which gained widespread acceptance in the mid-1970s after experimental confirmation of the existence of quarks. It describes the strong, weak, and electromagnetic fundamental interactions, using mediating gauge bosons. The species of gauge bosons are eight gluons, W−, W+ and Z bosons, and the photon.

The Standard Model also contains 24 fundamental fermions (12 particles and their associated anti-particles), which are the constituents of all matter.

(Excerpt end)

Observation:

The Standard Model has "widespread acceptance." It is impossible for anyone lacking a LHC to duplicate its experiments or to confirm any conclusions drawn from using the LHC.

Expensive projects like LHC serve a small community of certain scientists.

Quarks are not relevant to this book's use of fundamental particles (electron and proton) rather than their fragments.

The standard model offers no suitable explanation for a mass defect in atoms.

3.3 Structured Atomic Model

Recently, a new model for the atom has been proposed, called the structured Atomic Model, or SAM. This model presents a different explanation of an atom.

A video presentation is found by:
"Edwin Kaal: The Proton-Electron Atom — A Proposal for a Structured Atomic Model | EU2017"

Excerpt from its header:

Importantly, this model does not contradict the evidence in chemistry and physics, but rather provides a new interpretation and a promisingly fresh approach. With this model, Edwin has been able to resolve enigmas in chemistry and make predictions to inform future research.

(Excerpt end)

Excerpt from the Structured Atomic Model web site:

Scientists believe there are protons and neutrons inside the nucleus of the atom. However there is much disagreement as to how it may be organized or even if it has an organization. Quantum mechanics states we can only understand the nucleus through mathematics and that the location of nucleons can have a range of locations, we will never know of a structure. Is this a reasonable conclusion? We know there are particles in there, it stands to reason they are arranged in some manner. Is it possible we just haven't discovered it yet?

The Structured Atom Model is a theory that the nucleus is highly organized and stable. This organization is responsible for the properties of the elements and chemistry. The rules which define how the nucleus is built are relatively simple and adhere to key principles found in nature.

(Excerpt end)

Observation:

To put it simply, the nucleus in the structured atomic model consists of only protons and electrons. A neutron is a temporary bond between the 2. When this pair leaves the nucleus, the bond dissolves leaving the pair in a few minutes. This is a known behavior for a neutron.

The difference in this model is there can be other particle combinations within a nucleus, like the alpha particle. This structure enables an explanation for the alpha particle emission during radioactive decay. This decay step is awkward for a nucleus having a random arrangement of its nucleons, as implied in the Standard Model.

The structure of the protons and neutrons in the nucleus drives the properties of the atom, including the configuration of its electron orbitals.
However these orbitals really behave (their proposed shapes have evolved over time), absorption and emission lines are observed whenever electrons change their energy state within the atom.

This book does present an atomic model based on SAM.

The author started with a simple particle model of protons and electrons as the basis for how only protons and electrons can explain the observed measured mass of atomic matter, including a mass defect. SAM is never referenced again because it is not relevant to any explanations here. An alpha particle ejection is mentioned later, as well as the recognition that the order of particles being fused into a nucleus is currently unknown.

4 Planck's Constant

Planck's postulates were mentioned in the last section, with the Bohr model.

4.1 Bug in the equation

Another scientist recently concluded Planck's equation has a mistake in its units where h x f cannot = energy without a change.

This scientist had other conclusions as well, affecting subsequent sections in this book.

Planck's Constant and the Nature of Light

The YouTube video of that title by Lori Gardi is recommended for 5 reasons:

1) The well accepted Planck's equation has a bug. She has a thorough explanation which is worthwhile to hear.

A link to a web page of that title (the academic paper) is also in References.

A simple revelation worth noting here is:

Planck's equation has a mistake in its units. As a result:

$E = hf$ should be either:
a) $\Delta E = hf$
b) $E = htf$

where t is the time for the measurement.

She remarks this mistake and its fix have consequences for quantum mechanics.

2) The energy in light is in the intensity of a particular wave length, not only in the frequency, as is currently implied by the mistaken formula.

3) This video is another useful explanation of why there is no photon.

This book also concludes (in the next section), like other people, the Standard Model is not correct with its photon.

Light is a wave, not a particle. The electric and magnetic fields are oscillating with a consistent wave length during its propagation after initiation. Light never has a particle behavior. In an updated atomic model, there are no fictitious quasi-particles like a photon.

Particles require a mass to be detectable and measurable.

The conclusion that wave length intensity carries the energy is relevant to some absorption events.

4) Planck's constant defines the minimum wave length of light.

In some cases, the usage of Planck's constant must change because its units failed to address the missing time variable in Planck's equation. One usage is the uncertainty principle.

5) The uncertainty principle in quantum mechanics can have the uncertain limits defined so now they are not truly uncertain.

Copying much of her content here is not appropriate.

I have nothing to contribute to her excellent work. Excerpts can remove important context, in a case like this.

Observation:

From the video, this equation should be true:

$\Delta E = hc/\lambda$

Because a wave length is often used in this book, this particular formula is important.
The term "quantum of energy" for one photon refers to this equation. There is no photon, but a wave length can carry energy as calculated here.

4.2 The Reduced Planck's Constant

Planck's constant has a "reduced value" and its symbol is called h-bar.

Excerpt from Wikipedia:

The Planck constant has dimensions of physical action; i.e., energy multiplied by time, or momentum multiplied by distance, or angular momentum. In SI units, the Planck constant is expressed in joule-seconds (J*s or N·ms or kg·m^2/s). Implicit in the dimensions of the Planck constant is the fact that the SI unit of frequency, the hertz, represents one complete cycle, 360 degrees or 2π radians, per second. An angular frequency in radians per second is often more natural in mathematics and physics and many formulas use a reduced Planck constant (pronounced h-bar) -- thus apparently just J*s/cycle and J*s/radian units.

h = 6.626 07015 x 10^{-34} J·s

h-bar = h / 2π = 1.05405710817…x 1034 x10^{-34} J·s = 6.582 119 569… x10^{-16} eV·s

(Excerpt end)

Observation:

The h-bar value has seconds in its units, like h.

5 Photoelectric Effect

The photoelectric effect apparently resulted in the concept of a photon particle.

Excerpt from Wikipedia:

The photoelectric effect is the emission of electrons when electromagnetic radiation, such as light, hits a material. Electrons emitted in this manner are called photoelectrons.

The experimental results instead show that electrons are dislodged only when the light exceeds a certain frequency—regardless of the light's intensity or duration of exposure. Because a low-frequency beam at a high intensity could not build up the energy required to produce photoelectrons like it would have if light's energy was coming from a continuous wave, Albert Einstein proposed that a beam of light is not a wave propagating through space, but a collection of discrete wave packets, known as photons.

In 1905, Einstein proposed a theory of the photoelectric effect using a concept first put forward by Max Planck that light consists of tiny packets of energy known as photons or light quanta. Each packet carries hv energy that is proportional to the frequency v of the corresponding electromagnetic wave. The proportionality constant h has become known as the Planck constant.

The maximum kinetic energy K_{max} of the electrons that were delivered this much energy before being removed from their atomic binding is

$$K_{max} = h\nu - W$$

where W is the minimum energy required to remove an electron from the surface of the material.
Einstein's formula, however simple, explained all the phenomenology of the photoelectric effect, and had far-reaching consequences in the development of quantum mechanics.

(Excerpt end)

Observation:

Einstein described the phenomena but he did not justify a photon particle.

The energy requirement is defined by the atom.

The quantized behavior is in the atom, not in the light.

This is like a baby accepts only a mouthful of milk from the bottle. The amount in a mouthful is defined by the baby, not by the milk or the bottle.

Light is a continuous stream of energy not a collection of discrete wave packets.

Visible Light is a continuum of frequencies, essentially from violet to red. There are no discrete increments anywhere in this continuum of energy.

Our eyes see the combination of certain frequencies as white. Human eyes are not sensitive to only certain discrete packets.

There are no photons. Quantum mechanics just calls a wave length a photon.
However, wave lengths have no defined increment but span a continuum of values in whatever units are used, like Angstroms. The units selected for a measurement cannot define a behavior.

The photoelectric effect will be mentioned again in section 12 Atomic Equilibrium, placing it in context with other related behaviors.

6 Particle Pair Production

Excerpt from Wikipedia for particle pair production:

Pair production often refers specifically to a photon creating an electron–positron pair near a nucleus. For pair production to occur, the incoming energy of the photon must be above a threshold of at least the total rest mass energy of the two particles, and the situation must conserve both energy and momentum. However, all other conserved quantum numbers (angular momentum, electric charge, lepton number) of the produced particles must sum to zero – thus the created particles shall have opposite values of each other. For instance, if one particle has electric charge of +1 the other must have electric charge of −1, or if one particle has strangeness of +1 then another one must have strangeness of −1.

The probability of pair production in photon–matter interactions increases with photon energy and also increases approximately as the square of atomic number of the nearby atom.

(Excerpts end)

Observation:

Particle pair production probability increases with more electrons which matches the proton count.

Every atom has a specific configuration of its electron orbitals.

An atom will absorb a specific wavelength when it can change its set of electrons to a new energy state by that amount being absorbed. This is a quantized behavior of an atom, where a longer wavelength, having less energy than required, will not be absorbed by the atom.

Similarly, when an electron moves to a lower orbital, or to a lower energy state, an emission line of a particular wave length is emitted. This wave length is sometimes related to the distance between orbitals and it contains the energy being released from the electron's change in its energy.

The photoelectric effect has an extra result with the absorption line, by ejecting an electron.
When the atom absorbs enough energy for an electron to leave the atom rather than just changing orbitals, then the electron departs having the kinetic energy with the excess over the minimum required to leave.

The pair production event description is awkward in the excerpt, with "[creating a] pair near a nucleus" when describing the event is actually changing an electron pair in orbit around the nucleus. Of course, that orbit is "near."

Therefore, this is the proposed mechanism:

The gamma wave length energy is much greater than the ultraviolet wave length energy for a single electron ejection.

The substantial additional energy being absorbed by the atom from the energy in one short wave length causes another particle ejection, except the second electron flips its charge's negative polarity to positive becoming a positron.

Currently, particle pair production is claimed to create matter from energy.

With this alternate mechanism, there is no matter created during the event. Instead, an electron changed to a positron. The event caused no change in mass in any particles. Also, then there is no known mechanism to convert energy into matter, where matter is usually protons and electrons, the two components of every atom.

This alternate mechanism solves the antimatter problem. Physicists cannot explain the lack of antimatter in the universe. They assume with no justification, matter and antimatter should have been created in similar quantities.

Antimatter is actually created by only by high velocity particles, from either particle accelerators like the LHC, or from cosmic rays. The one exception is the positron, or anti-electron, which can occur during an atom's radioactive decay.

The antiproton is created when a proton flips its charge polarity, just as an electron can do, as described above. Currently, an antiproton can be created only in very high energy, high velocity particle collisions. This was done in 1955 but the specific particles or nuclei being used in the collision were not identified. Sometimes the deuteron, which is proton + neutron, is mentioned for particle colliders because there are several convenient sources for hard water, but never explicitly about an antiproton. Both participants in a collision must have protons, because electrons have so little mass for this. The simple explanation is protons are just flipping their charge polarity, to become an antiproton. This is more believable than an antiproton somehow appearing.

The presence of protons when an antiproton appears is not a coincidence. That is the only mechanism for creating an antiproton.

These mechanisms for anti-particles are infrequent in the universe so, of course, the anti-particles are rare.

No doubt, the problem of no antimatter arose from an assumption of what must follow the Big Bang. As I explained in my books, there was no big bang for the creation of the universe. That is almost religious dogma now.

The particle pair production event will be mentioned again in section 12 Atomic Equilibrium, placing it in context with other related behaviors.

7 Initial Changes to Atomic Model

The author started with a simple particle model, having no quarks, as the basis assuming only protons and electrons can explain the observed measured mass with atomic matter, including a mass defect. The quasi-particles would disappear along with the false photon. The simple beginning can be improved where necessary. There is much information on atomic behaviors, such as the Aufbrau principle. Such details are not relevant to a mass defect.
There are recommended changes for the Standard Model but the substantial evidence for certain known behaviors cannot be ignored.

The issue is not with valid science by evidence often obtained by a controlled experiment. The issue is with quasi-particles being proposed for behaviors having no need for such particles.

By the end of the book there are only a few significant changes recommended to the Standard Model.

The 4 notable ones are:

1) Reducing the mass of the proton,

2) Recognizing protons change their mass when compressed during fusion into a nucleus,

3) Dropping the quarks.
4) Dropping the photon and graviton,

Light is always a wave, never a particle. Light is oscillating electric and magnetic fields. The photoelectric effect is not a particle behavior.

A particle must have a mass. To verify a particle it must be measurable. Mass is measurable.

Light is measured as either a wave length or frequency (oscillations per time). These are wave behaviors, not particle.

Light is oscillating electric and magnetic fields and cannot posses mass.

Gravity is instantaneous and cannot be a particle requiring time to travel. Einstein was wrong about gravity limited to the velocity of light. Note: gravity never has a wave behavior. It cannot when it is instantaneous.

All the photon-related quasi-particles should be dropped.

The photon appears to be the basis of other quasi-particles. This might be because of Einstein's mistake about "information" having a velocity limit at c.

7.1 Definition of the atomic mass unit

The current atomic mass unit definition has a recommended change in this book.

This change can affect the claim of a mass defect, where the sum of the particles in an element does not add up to the element's measured atomic mass.

Some definitions from Wikipedia:

The dalton or unified atomic mass unit (symbols: Da or u) is a unit of mass widely used in physics and chemistry. It is defined as 1/12 of the mass of an unbound neutral atom of carbon-12 in its nuclear and electronic ground state and at rest. The atomic mass constant, denoted m_u, is defined identically, giving $m_u = m(^{12}C)/12 = 1$ Da.

By definition, the mass of an atom of carbon-12 is 12 daltons, which corresponds with the number of nucleons that it has (6 protons and 6 neutrons). However, the mass of an atomic-scale object is affected by the binding energy of the nucleons in its atomic nuclei, as well as the mass and binding energy of its electrons.
Therefore, this equality holds only for the carbon-12 atom in the stated conditions, and will vary for other substances. For example, the mass of one unbound atom of the common hydrogen isotope (hydrogen-1, protium) is 1.007825032241 Da, the mass of one free neutron is 1.00866491595 Da, and the mass of one hydrogen-2 (deuterium) atom is 2.014101778114 Da.

In general, the difference (mass defect) is less than 0.1%; exceptions include hydrogen-1 (about 0.8%), helium-3 (0.5%), lithium (0.25%) and beryllium (0.15%).

1 u or 1 Da = $1.66053906660 \times 10^{-27}$ kg
1 1 u = 1822.888486209 m_e

1 u = 1822.888486 m_e
m_p = proton mass = 1.007276466621 u

e = electric charge = $1.602176634 \times 10^{-19}$ C
proton charge = +1e
m_e = mass electron = $5.48579909070 \times 10^{-4}$ u
electron charge = -1e

neutrino mass = < 2.14×10^{-37} kg, 95% confidence level, sum of 3 flavors

neutrino charge = 0e

neutrino mass = < 3.53×10^{-10} u

(Excerpt end)

Observation:

Data came from several Wikipedia topics.

The neutrino mass is my calculation using the kg to dalton conversion. It should be in the list here because it is claimed to be a known subatomic particle though poorly understood, having an uncertain mass.

Excerpt from Wikipedia:

In physics, the proton-to-electron mass ratio, μ or β, is simply the rest mass of the proton (a baryon found in atoms) divided by that of the electron (a lepton found in atoms). Because this is a ratio of like-dimensioned physical quantities, it is a dimensionless quantity, a function of the dimensionless physical constants, and has numerical value independent of the system of units, namely:

$\mu = m_p/m_e = 1836.15267343$.

(Excerpt end)

Observation:

$1/\mu = 5.4462 \times 10^{-4}$

m_e uses this value and the proton mass.

At this point, the integrity of these assigned values could be checked.

However, the definition of 1 dalton does not provide the ^{12}C mass which was used for the calculation.

The particles in the ^{12}C atom, of 6 protons, 6 electrons, and 6 neutrons, which are each a proton and electron pair, can be summed with the result of 12.0873176 u

This is from:
12 times 1.007276466621 u for 6 protons and 6 neutrons
+
12 times $5.48579909070 \times 10^{-4}$ u for sum of 6 neutrons plus 6 pairs of a proton and orbiting electron.

Its current measured value in the Carbon isotopes topic is "exactly 12"
If the 12. value was actually used for calculating 1 Dalton then that use was a mistake. This isotope's mass value has the a mass defect.

The mass of protium, or ^{1}H, is provided and will be used for a better basis for calculating particle masses.

^{1}H = 1.007825032241

This atom is simply 2 particles:

^{1}H = m_p + m_e.

Using the two individual values the sum is
1.007825046530

My Excel value is slightly higher than from Wikipedia.

The current masses of an electron and proton do not add up to the mass in a protium atom.

There can be **no** other reason for this difference than the mass values, 1 or more of the 3 numbers involved, are wrong.

The protium (^1H) mass calculation can be from a different calculation using only one particle mass, not two:
$^1H = \mu\, m_e + m_e$ or $^1H = (\mu + 1)\, m_e$

This equation requires a high level of certainty of the precision of both the μ value and the ^1H value.

This result is 1.007825046538

This is not the measured value so either ^1H is wrong or m_e is wrong.

The m_e can be calculated with:

$m_e = {}^1H / (\mu + 1)$

with ^1H = 1.007825032240 (Excel fails with last digit as 1

This is spec: $m_e = 5.48579909070 \times 10^{-4}$ u

The new result is $m_e = 5.485799012873 \times 10^{-4}$

Calculation using 10 digits, $m_e = 0.0005485799$

Though the last digit is dropped for Excel, the result was slightly higher, but this is a debatable number of significant digits for a valid comparison.

This is not the current m_e so either ^1H is wrong or μ is wrong – or the current m_e is wrong. This topic proposes m_e must change.

The m_p can be calculated using m_e and the ^1H spec value:

$$m_p = {}^1H - m_e$$

With the calculated m_e value, m_p = 1.007276452331
Or 1.0072764523 with only 10 decimal digits

Compare with result from ^{12}C: 1.007276466621

This is the result with the new m_p and m_e ^1H= 1.007825032232
Compare with this spec value: ^1H = 1.007825032241

The 2 new values sum to a slightly lower ^1H by only at the limit of the software precision.

Both the old pair and the new pair add up to slightly less than the current atomic mass value, but beyond the significant digits.

My Excel 2003 handles up to 10 digits after the decimal point. The numbers add up with that precision.

I cannot define a new mass for an electron with a suitable number of significant digits, if more than 10 are required.

One could expect the community of people working with particle physics have a vested interest in agreeing on the correct values.

The atomic mass values for the elements are rarely, if ever specified with more than 10 digits after the decimal.

This book's goal is to be practical.

The current mass values are apparently wrong by a tiny amount. I expect there must be an agreement among many contributors for any change in mass of the 2 fundamental particles.

This book uses a value but if this recommendation is accepted, then a value with a defined precision, or the number of significant digits, must be agreed upon by those managing the "official" values.

This simple exercise using only ^1H and µ indicates physicists must confirm both values to the required precision, before assigning a mass value to the electron and proton if this alternate ^1H baseline is used instead of ^{12}C.

The precision of the two crucial input values affects the precision of the resulting particle mass values.

It is simply impossible for there to be "nuclear binding energy" in a nucleus consisting of only a proton.

Using the protium atom should be a better choice for defining the atomic mass unit because:

a) it consists of only the 2 fundamental particles,
b) it does not have 18 particles (6 x proton, 6 x electron, 6 x neutron) like ^{12}C,
c) it has no possible binding energy,
d) it is not clear how or whether binding energy is accommodated in the current ^{12}C algorithm,
e) it makes sense to use the unbreakable electron as the benchmark for defining atomic mass,

f) it is consistent with the updated atomic model treating the electron and proton as fundamental particles,

g) the issue with this selection is it requires an accurate proton-to-electron mass ratio,

h) the precision of m_e depends on the precision of only ^1H and μ

The recommendation is a change to ^1H for calculating masses should be considered "again."

The amu has a history worth noting about its element selection, described in this story:

Atomic Mass Unit Definition (AMU)

Excerpt:

John Dalton first suggested a means of expressing relative atomic mass in 1803. He proposed the use of hydrogen-1 (protium). Wilhelm Ostwald suggested that relative atomic mass would be better if expressed in terms of 1/16th the mass of oxygen. When the existence of isotopes was discovered in 1912 and isotopic oxygen in 1929, the definition based on oxygen became confusing.

Some scientists used an AMU based on the natural abundance of oxygen, while others used an AMU based on the oxygen-16 isotope. So, in 1961 the decision was made to use carbon-12 as the basis for the unit (to avoid any confusion with an oxygen-defined unit). The new unit was given the symbol u to replace amu, plus some scientists called the new unit a Dalton. However, u and Da were not universally adopted. Many scientists kept using the amu, just recognizing it was now based on carbon rather than oxygen.

At present, values expressed in u, AMU, amu, and Da all describe the exact same measure.

(Excerpt end)

Stating just "Wilhelm Ostwald suggested" does not provide his reason for it being "better." The subsequent discovery of isotopes indicated the selection probably was not better.

After trying protium first, then oxygen, then carbon, one can conclude protium should have remained the standard. It could be awkward to change the benchmark element for a third time, by a return to the initial choice.

A hydrogen-2 (deuterium) atom can be checked with the new particle masses because ^2H mass is provided. Its nucleus is a proton and neutron.

Measured: ^2H = 2.014101778114

$^2H = m_p + m_e + (m_p + m_e)$
Or it is twice ^1H
To expect: t 2.0156500644

This is more than the specified mass.

When using the old calculated values:

Measured ^2H = 2.015650093

The differences are tiny, but notable. There is the expected, non-quantified, binding energy between the 2 nucleons.

With new particle masses, the difference is -0.000000029

Therefore, there is a known mass defect with the deuterium atom using the current particle masses. This is with either pair of values of electron and proton mass, based on either ^{1}H or ^{12}C.

This is the expected result because the measured is less than the sum, resulting in a mass deficit, which is called nuclear binding energy. The proton is binding with the neutron.

This comparison has 2 alternate explanations:
1) a neutron exhibited a loss in mass.
2) a neutron exhibited a loss in its reactivity to other masses.

This book suggests the second. No mass is becoming energy.

The difference between sets for ^{1}H ^{12}C can be compared for their summation for ^{12}C.

^{12}C is measured at 12.0

Using the respective values for m_p and m_e, the results are:

From spec values ^{12}C = 12.09390056

From ^{1}H values ^{12}C = 12.0939004

There is a very small difference in the calculated values, beyond the number of significant digits (4 after decimal point) in the ^{12}C value.

However, the values which were supposed to result in exactly 12.0096 but their sum clearly failed to do so.

Both sets exhibit a mass defect because they do not match the measured atomic mass value,
The proton and electron mass values, derived from ^{12}C failed to result in exactly 12.0096. Because that was the goal of that algorithm, the algorithm failed. It did not account for nuclear binding energy.

That observation leads to a recommendation to use protium, which is the only atom having no possible binding energy.

Conclusion of atomic mass analysis:

This exercise suggests the current proton mass value is a tiny bit high.

The value derived from protium might be closer to the "correct" value if this is the only way to calculate the mass of a proton and electron.

Instead of recommending a change, the current value for the mass of an electron can remain unchanged, for now.

The big change in particle physics is explaining the observed mass defect.

Mass defect is currently explained as an awkward mass to energy conversion possibly suggested by Einstein. By the slight reduction in the proton mass, the mass defect in the protium atom is removed, as it should be, because the simplest nucleus has nothing to bind.

As a result, all elements will have a small reduction in their calculated mass defect.

Mass defect is detailed in the section Mass Defect Introduction.

7.2 A simpler subatomic particle model

Simpler means no quarks. This is the basis for the subsequent section about a new mechanism for gravity.

The few fundamental subatomic particles have a simple definition: mass and charge.

The current definition of atoms including the configuration of electron shells is unchanged. The only tiny change in the atom is in the mass of a proton.

There are only 2 fundamental subatomic particles. Their anti-particles arise infrequently when the particle has its charge polarity flipped in a high energy event.

1) electron
   mass = m_e, charge = -1e

2) proton
   mass = m_p, charge = +1e

Theneutrino, a possible third particle must remain tentative.

3) neutrino
   mass = non-zero, charge = 0e

Neutrinos do not carry a measured mass.

A neutrino is a questionable particle lacking certainty with its inconsistent detection.

The Standard Model claims there is an antineutrino.

The difference between neutrino and antineutrino is described by an attribute called chirality. Its only purpose is identifying its partner but nothing about itself, so it exists only to balance a particle equation, not as a real attribute which could be measured. Measurements provide evidence. When chirality is impossible to measure then it is meaningless.

An anti-particle is a flip in normal charge polarity, but a neutrino has no charge. There is no antineutrino in the simpler, updated atomic model.

In this list of 3, only 2 particles have mass. A particle must be measurable. Of the 3, there is only one particle, a neutrino, having no charge.

In a later section, the force of gravity is proposed as similar but distinct from an electric force which is driven by an electric charge.

Newton defined a simple behavior between masses, regardless of their charge.
The amount of mass in a particle determines how strongly it reacts to the presence of other masses.

Mass reactivity involves generating a field which other masses react to, just like charges react to other charges.

The strength of this effect of mass is reduced by free space, so the effect is weaker with increasing distance. In the terminology of an electric field, the density of the field lines diminishes with increasing distance from the source.

Later, the mass defect behavior appears driven by compression of nucleons.

The author's assumption of a proton as a fundamental particle is not affected by this observation.
The compression in size of a proton is causing a reduction in its reactivity. The reactivity to other masses is a particle behavior. There is no problem here.

It is reasonable to expect the particle's reactivity to other masses is in some way proportional to the particle's volume.

In non-technical terms, a fundamental particle has reactivity to other masses. Section 10 describes this behavior. A fundamental particle must possess the internal energy to generate a field for other masses to react with. Similarly, the particle must possess the energy to react to the field generated by other masses. The size of the particle probably defines the strength of any field it can generate.

The amount of charge in a particle determines how strongly it reacts to the presence of other charges.

A charge has different behaviors than a mass.

A moving charge creates what is called a magnetic field, with its strength determined by the amount of charge moving together.

A charge does not react to that magnetic field.

If the magnetic field is changing then it creates an electric field, so a charge will react to this change in its environment.

When a particle has the capability to react to other charges, every particle can have only one measured value of that reactivity, what is called the charge of an electron.

Having only one value of charge implies there is a single internal mechanism, among many particles, driving its reactivity to other charges. This mechanism is active or not in each particle, and is determined at the moment of the particles creation. The only particle where it is not active is the neutrino. A neutrino has a mass reactivity so tiny it is barely measurable.

Conversely, there is internal mechanism, among all particles, driving its reactivity to other masses. As noted above, the size of a particle affects its mass reactivity.

There has never been a measurement of a particle having a different amount of charge than others, so the charge behavior is apparently not affected by the particle's size.

As the fundamental particles are coherent, without individual components, this suggests there is an undefined minimum mass reactivity required in a particle before it is capable of charge reactivity.
When the particle has this reactivity to charges, it has only one level of reactivity but 1 of 2 possible states.

This state, or polarity, either "+" or" –" , is selected at the moment of particle creation.

The presence of anti-particles suggests the polarity behavior in a particle can be flipped to the other sign, without affecting the amount of charge reactivity.
Its charge polarity is either positive or negative. When positive, it will be repelled by a net positive in its environment or attracted to a net negative in its environment. When negative, the opposite reaction occurs. This reaction is based on the polarity of the electric field lines in its environment.

There is no measured change of the "charge of an electron" in any particle. The mechanism is consistent, though one can wonder how thoroughly such a value can be measured in individual particles.

7.3 Comparison of forces

Atoms have charged particles.
This is a comparison of those forces.
The neutron has 2 distinct particles of opposite charge, They are bound by both an electric force and a gravity force.

Values are from Wikipedia.

These equations are described again later.

Force of gravity equation:

$$F_g = G * (m1 * m2) / r^2$$

$G = 3.67430 \times 10^{-11} \: m^3 \cdot kg^{-1} \cdot s^{-2}$

e = electric charge = $1.602176634 \times 10^{-19}$ C

$m_e = 9.1 \times 10^{-31}$ kg

$m_p = 1.67262192369 \times 10^{-27}$ kg

$radius_e = 2.8179403227 \times 10^{-15}$ m

$radius_p = 0.84 \times 10^{-15}$ m

The electric force equation:

$$F_e = k_e * (q_1 * q_2) / r^2$$

$$k_e = 8.99 * 10^9 \ N \cdot m^2 \cdot C^{-2}$$

For the ^1H atom:

q_1 or $q_2 = e$
$m_1 = m_e$
$m_2 = m_p$

^1H atom radius = Bohr radius = $5.29 \times \times 10^{-11}$ m

So for proton and electron, in ^1H atom:

The gravity force:

$$F_g = 1.9795009170 \times \times 10^{-47} \ N$$

The electric force:

$$F_e = -8.2464899708 \times 10^{-8} \ N$$

F_e is negative meaning the force is attractive between opposite polarities.

When comparing F_e to F_g, the electric force is roughly 2.27×10^{39} stronger than the force of gravity.

The neutron will be mentioned later, but here is the appropriate place for this calculation and observation.

The neutron is usually in a nucleus. It disintegrates in a few minutes, when outside a nucleus.

So for proton and electron, in a neutron with $radius_p$ as their distance:

The gravity force: $F_g = 7.2468016237 \times \times 10^{-23}$ N

The electric force: $F_e = -1.8866878216 \times \times 10^{-17}$ N

Protons will be adjacent to other protons in an atom's nucleus.

Using double the proton radius as the distance between their centers, the electric force and the force of gravity can be calculated.

The gravity force:

$F_g = 1.9795009170 \times \times 10^{-47}$ N

The electric force:

$F_e = +81.7639597476$ N

F_e is positive meaning the force is repulsive between the same polarity (positive) particles.

The amount of force in F_e is substantial.

This large value should be noted, because an alpha particle ejection velocity will be described later.

The strong force within a nucleus must overcome this repulsion among the protons in near proximity, or even adjacent.
This repulsive force must be overcome during the process of fusion, where nucleons are added to a nucleus.

The electric and gravitation forces were described above, but in an atomic nucleus, the strong force must dominate over the other forces, because most nuclei have more than one proton which repels others.

7.4 Neutron

The neutron will be mentioned again in the mass defect section, but here is an appropriate place for this observation.

Excerpt from Wikipedia for Neutron:

The neutron is a subatomic particle, symbol n or n^0 which has a neutral (not positive or negative) charge and a mass slightly greater than that of a proton. Protons and neutrons constitute the nuclei of atoms. Since protons and neutrons behave similarly within the nucleus, and each has a mass of approximately one atomic mass unit, they are both referred to as nucleons. Their properties and interactions are described by nuclear physics.

(Excerpt end)

Observation:

This book assumes an atom consists of protons and neutrons in the nucleus with electrons in orbit around it. The existence of quarks is never part of this author's explanations.

A neutron is created by a tight bond between a proton and an adjacent electron, or in direct surface contact. They have an opposing charge.

The electric force between them was calculated earlier, using the proton radius as the distance between charges:

$F_e = -1.8866878216 \times \times 10^{-17}$ N

When a neutron is not in the nucleus, this force is not sufficient for a permanent bond. The disintegration into the original 2 particles within a few minutes is evidence of that.

However, when the neutron is compressed into a proton, or fused together, the result is deuterium or ^2H, then the electron is firmly attached, because ^2H does not decay. Doubling the mutual electrostatic force maintains the electron in the nucleus. Fusing another proton increases the positive charge in the nucleus and ^3He is stable so its neutrons remain intact.

The expected mass of a neutron is the sum of its 2 particles.

Neutron mass = $m_p + m_e$

Using current particle masses:

The result is 1.0078250465

The current measured mass of a neutron is 1.0086649159

The difference is 0.000839869350

Note the sum is greater than the measured.

The standard model accepts a neutron's measured mass is greater than the sum of its 2 particles.

This difference is called a mass defect. This is not a deficit, so there is no loss to binding energy.

Section 15 Periodic Table calculates the mass of a neutron in many isotopes and its value is never greater than expected from this sum.

However, the public value of a neutron's mass was measured indirectly using a deuterium atom so its accuracy could be questioned before making any conclusions about it.

Various nucleon combinations in a nucleus will be measured later, in the periodic table section.

The measured mass of a neutron will be calculated many times for many isotopes. The values are not consistent.

It is reasonable to conclude the decrease in reactivity to another mass, during a bond between particles, occurs in the proton, not the electron. The electron is somewhat attached to the much larger proton, while the protons are fused or compressed together within the nucleus.

This result indicates the 2 charge reactivity components of opposite polarity remain intact while observed as a neutron.

This explains its net charge of zero, and the separate polarities are maintained before and after disintegration.

Beta minus decay is a neutron decaying into a proton, electron, and electron neutrino.

However, this decay cannot be measured with precision.

According to Wikipedia:

The latter number is not well-enough measured to determine the comparatively tiny rest mass of the neutrino (which must in theory be subtracted from the maximal electron kinetic energy); furthermore, neutrino mass is constrained by many other methods.
A small fraction (about one in 1000) of free neutrons decay with the same products, but add an extra particle in the form of an emitted gamma ray.

This gamma ray may be thought of as a sort of "internal bremsstrahlung" that arises as the emitted beta particle (electron) interacts with the charge of the proton in an electromagnetic way. In this process, some of the decay energy is carried away as photon energy. Internal bremsstrahlung gamma ray production is also a minor feature of beta decays of bound neutrons, that is, those within a nucleus.

A very small minority of neutron decays (about four per million) are so-called "two-body (neutron) decays", in which a proton, electron and antineutrino are produced as usual, but the electron fails to gain the 13.6 eV necessary energy to escape the proton (the ionization energy of hydrogen), and therefore simply remains bound to it, as a neutral hydrogen atom (one of the "two bodies"). In this type of free neutron decay, in essence all of the neutron decay energy is carried off by the antineutrino (the other "body").

(Excerpt end)

Observation:

A gamma ray is energy in a wave length of radiation and is not a particle. Gamma ray generation is complex. Its generation during an alpha particle ejection is covered in section 12 Atomic Equilibrium. It is not critical to this decay behavior here , but the neutrino is important.
2) The origin of a neutrino is not fully understood. This beta decay explanation continues after this relevant excerpt.

Excerpt from Wikipedia:

Research is intense in the hunt to elucidate the essential nature of neutrinos, with aspirations of finding:

- the three neutrino mass values
- the degree of CP violation in the leptonic sector (which may lead to leptogenesis)

- evidence of physics which might break the Standard Model of particle physics, such as neutrinoless double beta decay, which would be evidence for violation of lepton number conservation.

(Excerpt end)

Observation:

Neutrinos are not well understood, but a beta decay results in one.

A neutron is an adjacent proton and electron. When they bind the resulting mass is unknown. It is impossible, despite many calculations for this book, to get a consistent value of a neutron because it can be measured only when fused and compressed in a nucleus. The excerpt above is for a free neutron or outside a nucleus. The topic for beta decay while in a nucleus never mentions the gamma ray.

The reason for including this beta decay description is the creation of this neutrino in this particular scenario is important. Both the proton and electron are cohesive bodies, with both capable of reacting to both charge and mass.

The proton, as implied from measurements done by others, can be slightly compressed. That indicates inherent structure and surface.

The 2 particles are held together by Coulomb's force between charges. When this physical connection between surfaces is broken, some how a particle possessing the mass behavior is created. Unfortunately, a measurement of its mass is never available.

One can imagine a behavior like the proton sheds a flake of its surface from the break between contact and this flake has enough material to be capable of reacting to other masses. Unfortunately, this is an unacceptable explanation when leading to deterioration of protons.

When reading descriptions of events involving neutrinos, it is almost impossible to determine whether a neutrino exists in a particle equation only to balance the quarks. A free neutron results from a step in the radioactive decay of isotopes with excess neutrons. If every free neutron emits a neutrino on its break-up, there could be many being emitted.
One could question whether they have been consistently detected to confirm the theory's prediction. When considering an atomic model with no quarks, a neutron is only a proton and electron, so the detection of these neutrinos in the description is important.

Neutrinos are described again in section 12 Atomic Equilibrium.

7.4.1 Neutron's mass

One might expect the mass of a neutron is measured directly. That is not the case. Its value uses the assumption of nuclear binding energy which this book suggests is not really as claimed.

Excerpt from Wikipedia:

The mass of a neutron cannot be directly determined by mass spectrometry due to lack of electric charge. However, since the masses of a proton and of a deuteron can be measured with a mass spectrometer, the mass of a neutron can be deduced by subtracting proton mass from deuteron mass, with the difference being the mass of the neutron plus the binding energy of deuterium (expressed as a positive emitted energy). The latter can be directly measured by measuring the single 0.7822 MeV gamma photon emitted when neutrons are captured by protons (this is exothermic and happens with zero-energy neutrons), plus the small recoil kinetic energy) of the deuteron (about 0.06% of the total energy). The energy of the gamma ray can be measured to high precision by X-ray diffraction techniques, as was first done by Bell and Elliot in 1948. The best modern (1986) values for neutron mass by this technique are provided by Greene, et al. These give a neutron mass of:
[mass] neutron = 1.008644904 u
The value for the neutron mass in MeV is less accurately known, due to less accuracy in the known conversion of u to MeV:
[mass] neutron = 939.56563 MeV/c^2.

Another method to determine the mass of a neutron starts from the beta decay of the neutron, when the momenta of the resulting proton and electron are measured.

(Excerpt end)

There are many assumptions to get the precise result, including a recognized issue with the "less accuracy" of the conversion from a gamma ray wave length to its equivalent mass.

The mass defect behavior appears driven by neutrons. The measured mass of a neutron was calculated by using the atom with a proton and neutron in its nucleus, so the conditions inherently involved a mass defect, which is a behavior lacking a good explanation in the standard model.

This technique is suspicious.

This brings into doubt whether the measured mass, defined with much precision, is actually verified to that accuracy.

Calculations using many significant digits must beware factors lacking the precision of other values when claiming the final precision.

The neutron behaves like 2 distinct particles but when in a nucleus, it can exhibit a reduced reactivity to other masses. This is the conclusion after the comparison between the 2 atoms having 3 nucleons but different sets.

The 2 distinct charge behaviors appear intact but the reactivity to other masses is changed slightly only while bonded.

When the split occurs, each charge components gets its original mass reactivity component.

The split is accompanied by a neutrino having little or no mass reactivity and no charge reactivity.

When an electron and proton unite to form a neutron, the mass reactivity, from the accepted value) is less than the sum.

This confirms the mass behavior is not just an addition. The mass reactivity of a neutron is not driven by the sum at its creation. In other words, a neutron can exhibit a difference in its expected mass while in a nucleus. This is called a mass defect. Mass defect is explained in the Periodic Table section in this book.

7.5 Nucleus

The size of some atomic nuclei has been measured.

Excerpt from Wikipedia:

The diameter of the nucleus is in the range of 1.7566 fm (1.7566×10^{-15} m) for hydrogen (the diameter of a single proton) to about 11.7142 fm for uranium. These dimensions are much smaller than the diameter of the atom itself (nucleus + electron cloud), by a factor of about 26,634 (uranium atomic radius is about 156 pm (156×10^{-12} m)) to about 60,250 (hydrogen atomic radius is about 52.92 pm). (Excerpt end)

Observation:

Uranium has from 235 to 238 nucleons.

The diameter change from 1 to 235 nucleons is from 1.7566 fm to 11.7142, or roughly a range of the size multiplier at about 7.

A study concluded it is difficult to measure the diameter of a nucleus and its distribution of the neutral neutrons. A link is in References.

Its title: Charge, neutron, and weak size of the atomic nucleus

The study included a measurement of the charge radius of the Calcium-40 and -48 nuclei.

Excerpt: "Our results for the charge radii are 3.49(3) fm for ^{40}Ca and 3.48(3) fm for ^{48}Ca"

Excerpt for Charge Radius:

The rms charge radius is a measure of the size of an atomic nucleus, particularly the proton distribution. It can be measured by the scattering of electrons by the nucleus. Relative changes in the mean squared nuclear charge distribution can be precisely measured with atomic spectroscopy.

The problem of defining a radius for the atomic nucleus is similar to that of defining a radius for the entire atom; neither atoms nor their nuclei have definite boundaries. However, the nucleus can be modeled as a sphere of positive charge for the interpretation of electron scattering experiments: because there is no definite boundary to the nucleus, the electrons "see" a range of cross-sections, for which a mean can be taken. The qualification of "rms" (for "root mean square") arises because it is the nuclear cross-section, proportional to the square of the radius, which is determining for electron scattering.

(Excerpt end)

Observation:

Apparently, adding 8 neutrons caused a tiny reduction of 0.01 fm in the charge radius, so the volume is not directly related to its particle count.

Noted above, comparing nucleons 1 to 235 resulted in diameter 7x

In the cited study, comparing nucleons 1 to 48 resulted in charge radius 4x, which is a not a similar ratio.

Protons and neutrons are considered of comparable size.

If the number of spheres in a volume is increasing faster than the volume, then there must be some compression of the spheres for that result.

Using the proton radius above, its volume is 8.78E-16 m^3
Using the Uranium diameter above, its radius is 5.87E-15 m for a volume of 8.48E-43 m^3

That volume can hold 298.7 protons except that quantity allows no space. Protons are spheres not a liquid which can fill available space with uniform density when gravity levels it.

Sphere packing is an awkward topic here when having no information about the arrangement of nucleons.

A random packing of equal spheres generally has a density around 64%.

That density value probably means the Uranium measured nucleus can hold 64% of the maximum number of protons, which results in a capacity of 191.

The reference does not identify which uranium isotope was measured. Their counts of nucleons are similar.

All would have the same result: the nucleons are being fused and compressed, resulting in a smaller volume than possible by random packing.
The calcium-40 and 48 nuclei can be checked here also.

The calcium-40 nucleus's nucleon capacity is 40.2 when packing at 64%.
Calcium-48 was measured smaller than ^{40}Ca and its capacity is 39.8.

One could expect the process of getting a nucleon added to a nucleus requires substantial force for the new object to adhere to the rest and that action could cause deformations to the joiner and perhaps to others already bound together.

Conditions in a stable nucleus are not clear.

The very few measured nuclei seem to have their nucleons densely packed suggesting no space between adjacent particles.

There is an unanswered question: when 2 protons are physically adjacent, is there an active Coulomb's force between them?

The combination of a stable of 3He and an unstable di-proton indicates a neutron is critical to nucleus stability. One must note adding more neutrons to a stable nucleus can make it unstable.

This is the case with $2H + n^0$ becoming 3H which has a half-life of 12.32 y.

The strong force is name usually assigned to the force maintaining the nucleons together but it has a poorly defined origin or mechanism.

Excerpt from Wikipedia:

In nuclear physics and particle physics, the strong interaction is the mechanism responsible for the strong nuclear force, and is one of the four known fundamental interactions, with the others being electromagnetism, the weak interaction, and gravitation. At the range of 10^{-15} m (1 femtometer), the strong force is approximately 137 times as strong as electromagnetism, a million times as strong as the weak interaction, and 10^{38} times as strong as gravitation.

In the context of atomic nuclei, the same strong interaction force (that binds quarks within a nucleon) also binds protons and neutrons together to form a nucleus. In this capacity it is called the nuclear force (or residual strong force). So the residuum from the strong interaction within protons and neutrons also binds nuclei together.

As such, the residual strong interaction obeys a distance-dependent behavior between nucleons that is quite different from that when it is acting to bind quarks within nucleons.

Additionally, distinctions exist in the binding energies of the nuclear force of nuclear fusion vs nuclear fission.

(Excerpt end)

Observation:

Atomic nuclei "binding energies" will be described in the section titled Mass Defect Introduction.

Mass defect is a phenomenon which arises in the atomic nucleus.

Other than hydrogen, every atom has at least one neutron.

A neutron is assumed to be a proton having an adjacent electron.

The description of the strong force (above) is similar to the nuclear force description.

Excerpt from Wikipedia:

The nuclear force (or nucleon–nucleon interaction or residual strong force) is a force that acts between the protons and neutrons of atoms.
Neutrons and protons, both nucleons, are affected by the nuclear force almost identically. Since protons have charge +1 e, they experience an electric force that tends to push them apart, but at short range the attractive nuclear force is strong enough to overcome the electromagnetic force. The nuclear force binds nucleons into atomic nuclei.
The nuclear force is powerfully attractive between nucleons at distances of about 1 femtometre (fm, or 1.0×10^{-15} metres), but it rapidly decreases to insignificance at distances beyond about 2.5 fm. At distances less than 0.7 fm, the nuclear force becomes repulsive.

This repulsive component is responsible for the physical size of nuclei, since the nucleons can come no closer than the force allows.

By comparison, the size of an atom, measured in angstroms (Å, or 1.0×10^{-10} m), is five orders of magnitude larger. The nuclear force is not simple, however, since it depends on the nucleon spins, has a tensor component, and may depend on the relative momentum of the nucleons.

(Excerpt end)

The undefined mechanism of the strong force could be explained like this:
The force of compression required to fuse together nucleons must overcome the force of repulsion between protons. Nucleons are observed to drop their volume so this result requires the particles are adjacent. The compression force is from one side and the force of reaction is from the other side. Together the sphere could reduce its volume.

At the moment the surfaces of the 2 positively charged particles touch, the direction of the mutual Coulomb's force changes from repulsion to attraction.

This force is strong enough to enable stability but not permanence. If the physical connection is disturbed, the mutual attraction can return to repulsion.

Radioactive decay can include neutron or alpha particle ejection from a nucleus.

With this explanation, both the strong and weak forces go away. Both are instances of Coulomb's force between positively charged protons.

The weak force is exerted the moment the mutual Coulomb's force changes direction to repulsion.

The strong and weak forces apparently need an updated description.

7.6 Rydberg Constant

The Rydberg constant is important in particle physics.

Excerpt from Wikipedia:

In spectroscopy, the Rydberg constant is a physical constant relating to the electromagnetic spectra of an atom. The constant first arose as an empirical fitting parameter in the Rydberg formula for the hydrogen spectral series, but Niels Bohr later showed that its value could be calculated from more fundamental constants via his Bohr model. As of 2018, [this constant] and electron spin g-factor are the most accurately measured physical constants.

The constant is expressed for either hydrogen as R_H or at the limit of infinite nuclear mass as R_∞. In either case, the constant is used to express the limiting value of the highest wavenumber (inverse wavelength) of any photon that can be emitted from an atom, or, alternatively, the wavenumber of the lowest-energy photon capable of ionizing an atom from its ground state. The hydrogen spectral series can be expressed simply in terms of the Rydberg constant for hydrogen RH and the Rydberg formula.

In atomic physics, Rydberg unit of energy, symbol Ry, corresponds to the energy of the photon whose wavenumber is the Rydberg constant, i.e. the ionization energy of the hydrogen atom in a simplified Bohr model.

(Excerpt end)

Observation:

The formula for the Rydberg constant uses only me not m_p.
In PPPB, m_p has a recommended reduction to make sure: $m_p + m_e$ = mass of ^1H.

There should be no justification for that simple equation of masses to be wrong when both ^1H and m_e are correct.

Apparently, that recommended change in m_p cannot affect the Rydberg constant.

However, the formula for the Rydberg constant for hydrogen uses both m_e and m_p.

This means R_H could be reduced, but the change in m_p had no change in value with the 10 digit precision of my MS Excel.

Maybe this consistency should not be a surprise. The new m_p mass was based on the well-established electron-proton mass ratio. R_H must have the same basis.

8 Einstein and Maxwell

Einstein wrongly assumed mass had a velocity limit at c, the velocity of light in a vacuum.

There are various claims of a speed limit for light and for mass.

A YouTube video is titled:
Why is the speed of light what it is? Maxwell equations visualized

This video clearly explains the velocity limit for light but makes a drastic mistake when mentioning Einstein. Einstein is mentioned separately below

From Wikipedia topic on Maxwell's equations:

Maxwell's equations explain how these [light] waves can physically propagate through space. The changing magnetic field creates a changing electric field through Faraday's law. In turn, that electric field creates a changing magnetic field through Maxwell's addition to Ampère's law. This perpetual cycle allows these waves, now known as electromagnetic radiation, to move through space at velocity c.

Relative permittivity is the factor by which the electric field between the charges is decreased relative to vacuum.

(Excerpt end)

There are also factors for relative permeability and magnetic susceptibility.

Observation:

The propagation of light is a self-propagating series of electric and magnetic fields. Its velocity is determined ONLY by the medium. An instantaneous change in medium causes an instantaneous change in propagation velocity. This transition is observed at the surface of glass or water.

Conclusion:

This propagation begins and continues at the same velocity regardless of any velocity of the source of its propagation.

All of the above is correct, regardless of anything Einstein claimed.

He had nothing to do with the constant velocity of light.

Some question whether a moving light source affects the velocity of its light. Einstein had thought experiments about that, when he concluded only the special observer might see its velocity change.

The velocity of light propagation is always defined by the medium.

The velocity and direction of a light source affects the energy in the light but not its rate of propagation.

If the light source is moving then it has kinetic energy.

In thermodynamics, energy cannot be lost or gained but only exchanged or transformed.

Around the sphere of radiated energy, wave lengths are reduced in the direction of travel or increased in the opposite direction. Each wave length change is determined by the velocity and direction at that point relative to c, the constant velocity of light in a vacuum. If the light source is in a medium, the z is the same because the medium affects the propagation not the initiation. Energy is maintained around the radiated sphere at the instant of emission and the propagation velocity of the light emission is not affected.

The velocity and direction of a moving light source has no effect on the velocity of its light.

Einstein did have something to do with the velocity of mass, a mistake.

He worked only with the context of a moving observer, the "special" observer in both "general" and "special" relativity..

Einstein's belief, of a velocity limit on mass, was shared by others in the 1800's, but has no justification in physics.

For many years, Einstein's unjustified belief has been refuted.

Excerpt from Wikipedia:

In 1993, Thomson et al. suggested that the (outer) jet of the quasar 3C 273 is nearly collinear to our line-of-sight. Superluminal motion of up to ~9.6c has been observed along the (inner) jet of this quasar.

Superluminal motion of up to 6c has been observed in the inner parts of the jet of M87. To explain this in terms of the "narrow-angle" model, the jet must be no more than 19° from our line-of-sight.

(Excerpt end)

Observation:

A plasmoid, like in 3C 273 or M87, holds substantial electromagnetic energy. The sustained force of a magnetic field on charged particles can result in velocities far faster than c and this has been measured many times.

The motion of mass is not a process dependent on the medium.

Motion is affected by the medium only with its friction on the surface of the mass in motion. Friction is an exchange of kinetic energy to thermal energy. Friction does not define a velocity limit.

Applying a force to a mass results in its acceleration. The force can be maintained for a specific time to achieve the desired velocity. This is a continuous transfer of energy from the source to kinetic energy.

This process of energy transfer is observed during every launch of a space probe. Power is the amount of force during a time. The force required is determined by the mass and the time required for the desired velocity.

The available power is determined by the fuel supply, which is "full" at the moment of launch. The number and design of the respective stages determines the final velocity of the final stage which has the lowest mass, where individual stages provide the required amount of power for the velocity of the remaining stages.
Increasing the initial power can increase the final velocity.

Once a mass is in motion it has kinetic energy. It must maintain that energy, so it remains in motion, until this energy is transferred like with friction. Friction is a transfer of kinetic energy into thermal energy.

As long as a force continues to transfer more energy into more kinetic energy, the velocity must increase. There is no velocity limit during this energy transfer.

Einstein wrote equations causing changes to the moving observer when near the velocity of light, including relativistic mass, so their possible velocity limit was set in math not physics. At $c = v$, the relativistic mass is a mass divided by zero which is either infinite or not allowed!

From Wikipedia:
"Oxford lecturer John Roche states that relativistic mass is not referenced in nuclear and particle physics, and that about 60% of authors writing about special relativity do not introduce it."

Observation:

This unverified prediction by relativity is widely ignored, while some claim all predictions by relativity were confirmed.

That claim of relativistic mass is unverified and should be retracted for many reasons. Other problems with relativity are described in later sections in this book.
Proposing mass has a velocity limit at c means when applying more force to a mass when nearing or at the velocity of c, energy is being lost rather than transferred to kinetic energy.

This loss is a violation of thermodynamics.

The practical velocity limit for a specific mass is set by the power required to reach that velocity.

The velocity limit on light is instantaneously set by the medium.

Light always propagates at a velocity set by the medium.

There is no defined velocity limit on mass.

9 Gravity

9.1 Newton's Laws of motion

Isaac Newton defined 3 laws of motion, which are not actually relevant to this book. They are included here to avoid confusing Newton's set of laws with Kepler's set of laws.

These laws from Wikipedia:

1)
In an inertial frame of reference, an object either remains at rest or continues to move at a constant velocity, unless acted upon by a force.

2)
In an inertial frame of reference, the vector sum of the forces F on an object is equal to the mass m of that object multiplied by the acceleration a of the object: $F = ma$. (It is assumed here that the mass m is constant.)

3)
When one body exerts a force on a second body, the second body simultaneously exerts a force equal in magnitude and opposite in direction on the first body.

Some also describe a fourth law which states that forces add up like vectors, that is, that forces obey the principle of superposition.

(Excerpt end)

9.2 Law of universal gravitation

Isaac Newton defined the law of universal gravitation, one of the main topics of this book.

Excerpt from Wikipedia:

Newton's law of universal gravitation is usually stated as that every particle attracts every other particle in the universe with a force that is directly proportional to the product of their masses and inversely proportional to the square of the distance between their centers. The publication of the theory has become known as the "first great unification", as it marked the unification of the previously described phenomena of gravity on Earth with known astronomical behaviors.
This is a general physical law derived from empirical observations by what Isaac Newton called inductive reasoning. It is a part of classical mechanics and was formulated in Newton's work Philosophiæ Naturalis Principia Mathematica ("the Principia"), first published on 5 July 1687. In today's language, the law states that every point mass attracts every other point mass by a force acting along the line intersecting the two points. The force is proportional to the product of the two masses, and inversely proportional to the square of the distance between them.

The equation for universal gravitation thus takes the form:

$$F = G * (m1 * m2) / r^2$$

where F is the gravitational force acting between two objects, m1 and m2 are the masses of the objects, r is the distance between the centers of their masses, and G is the gravitational constant.

(Excerpt end)

Observation:
That is Newton's force of gravity equation which has been used many times with success.

Wikipedia continued:

Newton's law has since been superseded by Albert Einstein's theory of general relativity, but it continues to be used as an excellent approximation of the effects of gravity in most applications. Relativity is required only when there is a need for extreme accuracy, or when dealing with very strong gravitational fields, such as those found near extremely massive and dense objects, or at small distances (such as Mercury's orbit around the Sun).

(Excerpt end)

Observation:

That claim of Newton being superseded by Einstein is just nonsense.

The claim relativity is required only for "extremely massive and dense objects" suggests a black hole which exists only in the theory of relativity and does not really exist. Newton's law cannot apply to fictional objects with impossible theories.

9.3 Slingshot trajectory

When space probes follow a slingshot trajectory past a distant planet to increase the velocity of that probe, the validity of Newton's force equation is verified every time.

The site universetoday had a page titled:
How do gravitational slingshots work?

When NASA calculates a trajectory of a space probe using another body to change the probe's velocity, it uses the force of gravity defined by Newton. Curvature of space-time is not used.

The web page noted above has a description of how NASA calculates a slingshot to execute a change in a probe's trajectory; a video is provided also. NASA has certainly demonstrated their technique with numerous successful missions.

The calculation of a slingshot involves these critical values:
a) the mass of the probe
b) the mass of the planet
c) the velocity of the probe
d) the velocity of the planet.

During the probe's approach there is the mutual force of gravity between the two bodies where the paths of both bodies are affected simultaneously. Obviously the probe with a rather small mass is affected much more than the planet.

These calculations are based on the simple Newton force equation.

As described above, mismatch between two bodies results in free fall acceleration of the much smaller body toward the much larger body.

The complete algorithm for a trajectory must account for a possible transition from the mutual force affecting the probe's vector to a closer proximity changing to a free fall behavior.

With probe's rapid fly-by, it probably never changes its vector to be toward the heavier body.

Relativity is based on space-time curvature by a gravitational field. The special observer's path is assumed to curve by the other body's gravitational field.

It is impossible to know whether anyone attempted to use the tensor equations to verify the path being predicted by curvature matched the path predicted by Newton's gravity equations, one ia for the mutual force, and the other is for possible free fall.
Curvature never involves the mass of the observer because the use of a gravitational field can ignore it, unlike the mutual force of gravity.

In relativity, though neither body is driven by a special observer, during a fly-by both bodies would be curving toward the other,
Curvature also never describes an affect on the body exerting its gravitational field which is affecting the special observer's path. Relativity is limited to only the special observer and their reference frame.

NASA never used relativity in its calculations for a slingshot trajectory. NASA does not use space-time curvature when a precisely calculated path is required.

Relativity assumed gravity had a velocity limit of c. NASA assumes gravity is instantaneous.

While not a disproof of relativity this application just shows relativity's space-time behavior would not enable a precise gravitational slingshot and was never used.

10 Mechanism for force of Gravity

10.1 Background

This section was adapted from this topic in the author's book, Redefining Gravity.

Isaac Newton was quoted as saying:

"You sometimes speak of gravity as essential and inherent to matter. Pray do not ascribe that notion to me, for the cause of gravity is what I do not pretend to know, and therefore would take more time to consider of it."

Excerpts are from Wikipedia.

When the text might have an uncertain source, then Observation is used, but when inserted too frequently it is distracting.

Mathematical Principles of Natural Philosophy is a work in three books by Isaac Newton, in Latin, first published 1687.

Newton defined the behavior of $F = ma$.

A definition of mass:

Mass is both a property of a physical body and a measure of its resistance to acceleration (a change in its state of motion) when a net force is applied.
An object's mass also determines the strength of its gravitational attraction to other bodies.

With the publication of "A Dynamical Theory of the Electromagnetic Field" in 1865, [James Clerk] Maxwell demonstrated that electric and magnetic fields travel through space as waves moving at the speed of light.

(Excerpts end)

Observation:

Isaac Newton (178 years earlier) could not know how an electric field works so he could not propose a gravity field.

10.2 A New Mechanism for the Force of gravity.

This explanation begins with examining Maxwell's equations.

The video abut Maxwell's equations, noted in a previous section, offers a reader a visual presentation, which could supplement the text below about those equations.

There is at least one other theory of gravity proposing it is based on the electric force between charges in the subatomic particles in the respective masses.
Such a theory assumes there is only an electric force and gravity is just a manifestation of it, enabling a bipolar behavior.

This proposes gravity is a separate attractive force being dependent on the medium, just like the electric force.

Everyone knows the 2 inverse-square forces are similar in their equation format but there is a notable difference between them: gravity is much weaker.

When there are 2 similar mutual forces with similar behaviors but one is much weaker, then probably each force is uniquely affected by the medium.

The following explanation omits the calculus but hopefully this has enough detail.

Maxwell's equations define several properties of "free space" and those values define the rate of propagation of light through that free space.

Now, they can be considered properties of the medium, aka, the aether, which is whatever unknown "stuff" permeates the universe.

The medium defines the rate of propagation of the synchronized electric and magnetic fields within light.

Most know light travels slower through glass or water than through air or space.

The diffraction index is the factor defining the change in light velocity by the medium.

Essentially, the medium has a measurable resistance to the changing of electric and magnetic fields. During the propagation of light, both fields are oscillating or in continuous change.

Light is more complicated then that simple statement because different wave lengths have different behaviors like X-rays which can be either penetrating or shielded by different media. The color violet is slower than red through a glass prism.

At the foundation of Maxwell's equations are 2 constants which define how the medium affects changes in an electric field or a magnetic field:

the permittivity of free space, ε_0, epsilon-nought
the permeability of free space, μ, mu

These factors become Coulomb's constant.

The Electric force is described by Coulomb's law.

$$F = k_e * (q_1 * q_2) / r^2$$

where k_e is Coulomb's constant ($k_e \approx 8.99 \times 10^9$ $N \cdot m^2 \cdot C^{-2}$), q_1 and q_2 are the signed magnitudes of the charges, and the scalar r is the distance between the charges. The force of the interaction between the charges is attractive if the charges have opposite signs (i.e., F is negative) and repulsive if like-signed (i.e., F is positive).

In very simple terms, there is a force between any 2 charges.

This electric force is reduced by 2 factors:

1) k_e from the medium,

2) r from the distance.

Observation:

The units of k_e are essentially a ratio of force in an area relative to charge.

Free space defines a factor within ke resulting in a force reduction between charges.

After noting the role of free space in electromagnetism, the force of gravity is considered next.

The force of gravity is defined by Newton's Law of Universal Gravitation.

$$F = G * (m1 * m2) / r^2$$

where F is the gravitational force acting between two objects, m1 and m2 are the masses of the objects, r is the distance between the centers of their masses, and G is the gravitational constant.

The measured value of the gravitational constant is approximately $6.674 \times 10^{-11} \cdot m^3 / kg \cdot s^{-2}$

One might notice the mix of units in the force equation. There are seconds in its constant's units but there is no time variable in the factor value it multiplies, which has only these units: kg^2 / m^2

The units of force are $kg \cdot m \cdot s^{-2}$

This mismatch of time in units is reminiscent of Planck's equation, noted in section 4.

Wikipedia has a topic Cavendish Experiment describing how G was initially calculated using the oscillation of a torsion bar in an experiment taking 20 minutes. Its current accepted value is by measurement, not by a calculation using defined "free space" parameters.

I assume the units of ke are just a remnant from measurements during experiments.
Dropping the s^{-2} units from the constant (again, there is no time value in the equation) leaves only:

m^3 /kg

This factor, which looks like inverse density, defines a ratio between:

a) a distance (though this m^3 in the numerator implies a distance is being treated as a volume) and

b) a participating mass, with kg in denominator.

This factor is essentially a ratio of a distance per mass.

The multiplication results in less force per kg, because the G value is much < 1

Instead of "Gravitational constant" this factor could be named "Gravitational Gradient" because the force reduces over distance, based on the medium, though the underlying parameters are not identified as Maxwell did for an electromagnetic constant.

This ratio might be considered a "free space" behavior for a gravity field.

Electric and magnetic fields required individual free space parameters. A gravity field apparently requires its own free space parameter.

One goal of this book is: Defining a New Mechanism for Force of Gravity

Newton did not propose a mechanism for mass to drive its force of gravity.

The instantaneous force of gravity is the result of a mass field around every proton and electron.
Both particles already have an accepted electric field.

Both particles and atoms behave as expected for combinations of charges.

The 2 fields around both fundamental particles are different though the resulting mutual force affects both participants similarly.

The electric field is either attractive or repulsive while the gravity field is only attractive, but both are mutual.

This simple observation means all of Maxwell's equations for a static electric field and its mutual force also apply to this static gravity field and its mutual force.

Gravity is not electrical so permittivity for a capacitance in free space does not apply.

For gravity, open space is just a "resistance" to the force. That word is used in the definition of mass. The distributed free space resistance for this particular force explains why the force of gravity is so different between the masses of proton and electron compared to the force between their charges.

This new gravity field is NOT the accepted gravitational field around a sphere of uniform density causing free fall acceleration to smaller bodies near its surface.

Calling it a gravity field compared to an electric field is appropriate when one applies Maxwell's field equations.

The following is the Wikipedia description of an electric field but mixed with changes for an application to a gravity field highlighted.

I hope this from/to approach for changing a description from an electric field to a gravity field is clear. The exercise should reveal their similar behavior, to support this hypothesis.

1 From:

An electric field (sometimes E-field is the physical field that surrounds each electric charge and exerts force on all other charges in the field, either attracting or repelling them. Electric fields originate from electric charges, or from time-varying magnetic fields. Electric fields and magnetic fields are both manifestations of the electromagnetic force, one of the four fundamental forces (or interactions) of nature.

1 To:

A **gravity** field (sometimes **G-field** is the physical field that surrounds each **mass** and exerts force on all other **masses** in the field, attracting them. **Gravity** fields originate from **masses**. **Gravity** fields are manifestations of one of the four fundamental forces (or interactions) of nature.

G-field is used, not M-field, to avoid confusion with a magnetic field.

2 From:

The electric field is defined mathematically as a vector field that associates to each point in space the (electrostatic or Coulomb) force per unit of charge exerted on an infinitesimal positive test charge at rest at that point. The derived SI units for the electric field are volts per meter (V/m), exactly equivalent to newtons per coulomb (N/C).

2 To:

The **gravity** field is defined mathematically as a vector field that associates to each point in space the (force per unit of **kg** exerted on an infinitesimal positive test **mass** at rest at that point. The derived SI units for the **gravity** field are N per meter (**N**/m), exactly equivalent to newtons per kg (N/kg).

Image and caption (**3**) from Wikipedia Electric Field:

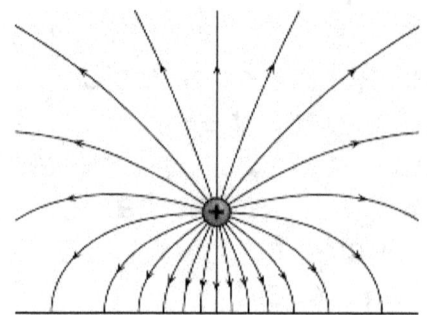

3 From:

Electric field of a positive point charge suspended over an infinite sheet of conducting material. The field is depicted by electric field lines, lines which follow the direction of the electric field in space.

3 To:

Gravity field of a point **mass** suspended over an infinite **span** of **gravity** conducting medium. The field is depicted by **gravity** field lines, lines which follow the direction of the **gravity** field in space.

4 From:

The electric field is defined at each point in space as the force (per unit charge) that would be experienced by a vanishingly small positive test charge if held at that point. As the electric field is defined in terms of force, and force is a vector (i.e. having
both magnitude and direction), it follows that an electric field is a vector field. Vector fields of this form are sometimes referred to as force fields. The electric field acts between two charges similarly to the way
the gravitational field acts between two masses, as they both obey an inverse-square law with distance.

4 To:

The **gravity** field is defined at each point in space as the force (per unit **mass**) that would be experienced by a vanishingly small positive test **mass** if held at that point. As the **gravity** field is defined in terms of force, and force is a vector (i.e. having both magnitude and direction), it follows that a **gravity** field is a vector field. Vector fields of this form are sometimes referred to as force fields. The **gravity** field acts between two **masses** similarly to the way the gravitational field acts between two masses, as they both obey an inverse-square law with distance. This is the basis for **Newton's** law, which states that, for stationary **masses**, the **gravity** field varies with the source **mass** and varies inversely with the square of the distance from the source. This means that if the source **mass** were doubled, the electric field would double, and if you move twice as far away from the source, the field at that point would be only one-quarter its original strength.

5 From:

This is the basis for Coulomb's law, which states that, for stationary charges, the electric field varies with the source charge and varies inversely with the square of the distance from the source. This means that if the source charge were doubled, the electric field would double, and if you move twice as far away from the source, the field at that point would be only one-quarter its original strength.

5 To:

This is the basis for **Newton's** law, which states that, for stationary **masses**, the **gravity** field varies with the source charge and varies inversely with the square of the distance from the source. This means that if the source **mass** were doubled, the **gravity** field would double, and if you move twice as far away from the source, the field at that point would be only one-quarter its original strength.

6 From:

The electric field can be visualized with a set of lines whose direction at each point is the same as the field's, a concept introduced by Michael Faraday, whose term 'lines of force' is still sometimes used. This illustration has the useful property that the field's strength is proportional to the density of the lines. The field lines are the paths that a point positive charge would follow as it is forced to move within the field, similar to trajectories that masses follow within a gravitational field.

6 To:

The **gravity** field can be visualized with a set of lines whose direction at each point is the same as the field's, a concept introduced by Michael Faraday, whose term 'lines of force' is still sometimes used. This illustration has the useful property that the field's strength is proportional to the density of the lines. The field lines are the paths that a point **mass** would follow as it is forced to move within the field, similar to trajectories that masses follow within a gravitational field. Field lines due to stationary **masses** have several important properties, including always originating from **point masses** and terminating at other **masses**, they enter all **masses** at right angles, and they never cross or close in on themselves. The field lines are a representative concept; the field actually permeates all the intervening space between the lines. More or fewer lines may be drawn depending on the precision to which it is desired to represent the field.

7 From:

Field lines due to stationary charges have several important properties, including always originating from positive charges and terminating at negative charges, they enter all good conductors at right angles, and they never cross or close in on themselves. The field lines are a representative concept; the field actually permeates all the intervening space between the lines. More or fewer lines may be drawn depending on the precision to which it is desired to represent the field.

7 To:

Field lines due to stationary **masses** have several important properties, including always originating from **masses** and they never cross or close in on themselves. The field lines are a representative concept; the field actually permeates all the intervening space between the lines. More or fewer lines may be drawn depending on the precision to which it is desired to represent the field.

8 From:

$$E(x_0) = F / q_0 = q_1 / (X_1-X_0)^2 ..$$

This is the electric field at point x_0 due to the point charge q_1; it is a vector-valued function equal to the Coulomb force per unit charge that a positive point charge would experience at the position x_0. Since this formula gives the electric field magnitude and direction at any point x_0 in space (except at the location of the charge itself, x_1, where it becomes infinite) it defines a vector field. From the above formula it can be seen that the electric field due to a point charge is everywhere directed away from the charge if it is positive, and toward the charge if it is negative, and its magnitude decreases with the inverse square of the distance from the charge.

The Coulomb force on a charge of magnitude q at any point in space is equal to the product of the charge and the electric field at that point

8 To:

$$G(x_0) = F / m_0 = m_1 / (X_1-X_0)^2 \ldots$$

This is the **gravity** field at point x_0 due to the point **mass** m_1; it is a vector-valued function equal to the **gravity** force per unit mass that a point **mass** would experience at the position x_0. Since this formula gives the **gravity** field magnitude and direction at any point x_0 in space (except at the location of the **mass** itself, x_1, where it becomes infinite) it defines a vector field. From the above formula it can be seen that the **gravity** field due to a point **mass** is everywhere directed away from the mass, and its magnitude decreases with the inverse square of the distance from the **mass**.

The **gravity** force on a charge of magnitude q at any point in space is equal to the product of the **mass** and the electric field at that point

End of the from/to sequence of 8 steps.

Observation:

The Wikipedia descriptions of an electric field behavior frequently have a reference to a similar behavior in gravity.

In some cases, the "To" text needed no change because gravity was already there.

Excerpt from Wikipedia Electric Field:

Coulomb's law, which describes the interaction of electric charges: is similar to Newton's law of universal gravitation:

This suggests similarities between the electric field E and the gravitational field g, or their associated potentials. Mass is sometimes called "gravitational charge". Electrostatic and gravitational forces both are central, conservative and obey an inverse-square law.

Each charge field is diminishing with distance. Their mutual interaction results in a mutual force.

(Excerpt end)

Observation:

A gravity field from a mass behaves the same with another mass like a pair of charges. The difference is an electric field has polarity and interacts with only other electric fields, or with a magnetic field.

A gravity field interacts with only other gravity fields, and is not affected by an electric or magnetic field.

The force of gravity, between gravity fields of its participants being pervasive, is instantaneous and does not propagate.

Gravity field is also simpler. Changing electric and magnetic fields create the other. A changing a mass cannot create another type of field.

The universe has pervasive charge fields and gravity fields, with "lines" to describe their relative strength.

This theory does not change Newton's force of gravity equation, which has been verified numerous times. It only tries to explain its mechanism.

Relativity broke Newton's valid application of the force of gravity. Relativity must be dropped by physics because Newton's force remains valid. Wikipedia claims relativity superseded Newton's force, which is such an incredible mistake. Even more so, when one realizes relativity applies only to a special moving observer, becoming quite irrelevant to sciences like cosmology when there is no special observer.
Any unified field theory having no special observer must ignore relativity.

This rudimentary theory is offered just because it seems either relativity or dark matter are always involved in every discussion of gravity.
Both are invalid and must be ignored when redefining the force of gravity.

Gravity has complex behaviors like orbital resonances.

This book is seeking a mechanism, not defining the math for all the behaviors.

An alternative proposed by some individuals, where gravity has a bipolar behavior, or both attractive and repulsive, is more complex than a gravity field.
This proposed theory should be more productive than some other theories having an inconsistent behavior, because gravity has a consistent behavior through many experiments and accurate slingshot trajectories.
I suspect the free space parameter for gravity has a different origin. "free space" remains a mystery to physics.

I have an unrelated observation, which must be stated.

Free space parameters, like for the electric and magnetic fields, are defined by the universal medium, or the aether. It is impossible to know whether these values are consistent throughout the entire universe.

11 Light

This section is adapted from this topic in the author's book, Observing Our Universe.

That book offers much more detail than necessary for this book about the atomic model. Only the basics are here.

11.1 Light and wavelengths

A spectrum is the entire range of wavelengths in electromagnetic radiation where light is the visible range. The ultraviolet and infrared ranges are not visible to the human eye but they are in the Sun's radiation. Because this topic is about the visible stars and galaxies, the word light is often used for the entire spectrum, including those frequency ranges not visible.

Electromagnetic radiation is the propagation of synchronized, perpendicular electric and magnetic fields. The propagation has a defined rate of oscillation measured as either a frequency or a wavelength. The wavelength is usually measured in either nanometers (10^{-9} m) or Angstroms (10^{-10} m or 0.1 nm). The velocity of this propagation has been measured in a vacuum using our standard definition for time and this measured value is called the constant c. This measurement also defined the standard unit of 1 meter. The velocity of propagation is reduced in a medium, defined by the medium's diffraction index. Light transmits energy proportional to its frequency so the constant c appears in some physics equations involving energy.

Quantum physics defined a theoretical particle called a photon to refer to a single wavelength.
In this section, wavelength is used because a spectrum analysis uses specific numerical values. Using photon instead of wavelength only introduces possible confusion.

11.2 Fraunhofer Lines

This description provides background for many terms and their use in a spectrum analysis.

Excerpt from Wikipedia:

In 1814, Fraunhofer independently rediscovered the [dark] lines and began to systematically study and measure the wavelengths where these features are observed. He mapped over 570 lines.

About 45 years later Kirchhoff and Bunsen noticed that several Fraunhofer lines coincide with characteristic emission lines identified in the spectra of heated elements. It was correctly deduced that dark lines in the solar spectrum are caused by absorption by chemical elements in the solar atmosphere. Some of the observed features were identified as telluric lines originating from absorption by oxygen molecules in the Earth's atmosphere.

Because of their well-defined wavelengths, Fraunhofer lines are often used to characterize the refractive index and dispersion properties of optical materials.

(Excerpt end)

11.3 Atom's characteristic wavelengths

Calcium and hydrogen are the most frequently observed atoms in the spectrum of a distant galaxy or quasar.

The calcium atom is important because a galaxy can have its ion's pair of calcium absorption lines at 3934 and 3969 Angstroms in its spectrum when a calcium ion is in the line of sight to the galaxy. A red or blue shift of this pair of lines indicates the relative velocity of the ion. The neutral calcium atom has a different pair of wavelengths.
Nearly all matter in the universe is plasma, or it has an electrical charge. That includes electrons (-), protons(+), and ions (+) which are atoms having lost one or more electrons.

Hydrogen is the most common element in the universe; it is also the simplest having only one proton and one electron.

Excerpt from Wikipedia:

In physics, the Lyman-alpha line is a spectral line of hydrogen, or more generally of one-electron ions, in the Lyman series, emitted when the electron falls from the n = 2 orbital to the n = 1 orbital, where n is the principal quantum number. In hydrogen, its wavelength of 1215.67 angstroms corresponding to frequency of 10^{15} hertz, places the Lyman-alpha line in the ultraviolet part of the electromagnetic spectrum, which is absorbed by air. Lyman-alpha astronomy must therefore ordinarily be carried out by satellite-borne instruments, except for extremely distant sources whose red shifts allow the hydrogen line to penetrate the atmosphere.

(Excerpt end)

Observation:

This wavelength is important because a quasar usually has this emission line in its spectrum. A shift of this emission line wavelength indicates the relative velocity of the atom.

11.4 Doppler Effect

Excerpt from Britannica:

Doppler effect, the apparent difference between the frequency at which sound or light waves leave a source and that at which they reach an observer, caused by relative motion of the observer and the wave source. This phenomenon is used in astronomical measurements.

(Excerpt end)

The Doppler Effect is observed by the entire spectrum of the light source being shifted in proportion to the source's velocity in that direction.
The velocity of light is set by the medium. The velocity of light cannot be affected by the light source velocity. However, the source in motion affects the distribution of the radiated energy, not its velocity.

The timing of the Doppler Effect is crucial when one observes a spectrum shift in radiation from distant objects.

The Doppler Effect occurs only at the moment of radiation emission, when the motion of the object at that instant affects the spectrum.

There are 2 sources of electromagnetic radiation affected by the Doppler Effect: stars and atoms. Each initiates the propagation of the synchronized electric and magnetic fields. This propagation is an expanding sphere from the source. This sphere of energy continues until it is absorbed by an object in its path.

Stars emit a broad spectrum of thermal radiation.

Atoms emit a characteristic wave length based on the electron configuration.

The energy being lost in the atom is transferred to the corresponding wave lengths of electromagnetic radiation. Some atoms emit more than one wave length when dropping to their ground state.

These wave lengths can be observed and measured in a spectrum, and are called emission lines.

The instant of radiation emission, the motion of the source affects the wave length distribution around that sphere. Wave lengths in the direction of the source are changed by an amount proportional to the sources velocity relative to the velocity of light. The light source is generating a continuum of energy as a sphere. Wave lengths in one side of the sphere will be reduced, or toward the blue end, in the direction of the source. Wave lengths in the other side of the sphere will be increased, or toward the red end, in the direction opposite of the source. There is perfect symmetry with the change in wave length on one side exactly matched by the change on the opposite side. The sphere is a continuum of energy, being carried in wave lengths. There is definitely no quantized behavior present.

The motion of the light source does not change the amount of energy being radiated, only its distribution around the sphere of its propagation. Energy is always conserved.

The Doppler Effect also occurs only at the moment of radiation absorption, when the motion of the object at that instant also affects the spectrum. When energy is absorbed by an object than that energy is missing from the radiation. The energy is carried in wave lengths so those wave lengths carrying the energy which was transferred to the object are missing in the spectrum. These missing wave lengths are called absorption lines.

Absorption lines arise from objects in the line of sight, between the light source which emits the intact energy or spectrum.

The absorption line behavior is affected by the velocity of the atom. A moving atom carries kinetic energy and that energy participates in the transfer of energy from the radiation to the atom. As with an emission line, the velocity of the atom relative to the velocity of light determines the energy involved.

An atom is essentially a tiny sphere. An atom in the path of electromagnetic radiation can absorb energy from that energy. The atom's motion relative to the radiation is important. The motion at that point in the sphere will have a proportion relative to the velocity of light and relative to the direction of the incoming light.

An atom has specific behaviors to execute when absorbing energy; these are defined in the Atomic Equilibrium section.

When the atom is moving toward the light source the kinetic energy of the atom is a participant and it reduces the energy the atom requires and absorbs from the radiation. This decrease in energy is a higher wave length.

Energy is always conserved during this exchange.

When the atom is moving away from the light source the kinetic energy of the atom is a participant and it increases the energy the atom requires and absorbs from the radiation. This increase in energy is a lower wave length.
The energy being absorbed is noted as an absorption line wave length.

The following is adapted from the author's book Observing Our Universe:

This is the simple calculation of z.

The velocity, called v here, of the source is compared to the velocity of light by dividing that value by the velocity of light, called the constant c.

The value of v has a sign. Doppler Effect is in the observer's line of sight. When the object is moving away from the observer, v is + or positive, and when moving toward the observer, v is − or negative.

The result is called z by convention.

The simple equation is $z=v/c$, making sure the units are the same (usually km/s).

The shift in a spectrum due to the motion of the light source is a simple equation,
where EWL is the emission wavelength,

NWL is the new wavelength, so:
NWL = EWL + (EWL multiplied by z)

where the z is the factor for the change in the new wavelength from that originally emitted; z is positive for a red shift or negative for a blue shift.

There is no quantized behavior in any of the equation's factors or in the result.

11.4.1 Galaxy Red Shift

This is adapted from the author's book Observing Our Universe.

The spectrum of galaxies beyond our Local Group exhibit a unique behavior. In 1936, Edwin Hubble noticed this and put our Local Group on an island separate from the Hubble Flow.

These galaxies have an absorption line which shifts toward the red, and this shift is roughly proportional to the galaxy's distance from the observer, who is always on or near the Earth. This single line was attributed to hydrogen.

The explanation for the line shift is hydrogen atoms in the intergalactic medium (IGM), the space beyond our Local Group of galaxies. These atoms absorb its wave length, but then drop to their ground state and re-emit the wave length again.

This process of absorption and re-emission results in this observed increasing red shift based on the density of hydrogen atoms in the IGM. This red shift is caused by the IGM in the line of sight, not the light source.

By mistake, this hydrogen absorption line red shift was considered the result of a velocity causing a Doppler effect. This is only a line of sight behavior and indicates nothing about the distant galaxy's actual velocity. This mistake caused many others, including the universe expansion, dark energy, and the Big Bang.

Improving imaging technology enables a spectrum to be captured from galaxies which had been too dim by their distance.
Because this galaxy red shift is driven by the distance through the IGM, their red shift also increases by this measurement. Essentially, the only limit on a galaxy red shift is the technology to measure the most distant ones. Treating this z as a velocity of the galaxy is ridiculous. As noted in the book, scientists eventually tried to explain how galaxies could possibly have a velocity exceeding 8x the velocity of light. Their conclusion was the red shift must be coming from the IGM.

11.4.2 Quasar Red Shift

This is adapted from the author's book Observing Our Universe.

A quasar is a distant object which looks like a star but it has a strong source of synchrotron radiation, extending from radio to X-ray, All share red-shifted emission lines from a variety of non-hydrogen elements where the mix can vary by quasar. All share the same red shifted hydrogen Lyman-alpha emission line.

These quasars can have this line with a red shift indicating the atom is moving at many multiples of the speed of light, like 7x. A proton when capturing an electron emits this wave length. The wave length is shifted by the proton's velocity at the instant of that capture. This red shift comes from the atom in the line of sight, and indicates nothing about the distant quasar's actual velocity. This mistake compounds the galaxy red shift mistake, so both objects having a different mechanism make the false dark energy difficult to explain both false velocities.

Also, a quasar's hydrogen red shift of $z > 1$ indicates a proton's velocity is exceeding that of light. Einstein developed the theory of relativity assuming mass cannot travel faster than c. His assumption was shown to be a mistake by many quasars. Relativity has too many mistakes.

11.5 Synchrotron Radiation

Excerpt from Wikipedia:

Synchrotron radiation, electromagnetic energy emitted by charged particles (e.g., electrons and ions) that are moving at speeds close to that of light when their paths are altered, as by a magnetic field. It is so called because particles moving at such speeds in a variety of particle accelerator that is known as a synchrotron produce electromagnetic radiation of this sort.

Many kinds of astronomical objects have been found to emit synchrotron radiation as well. High-energy electrons spiraling through the lines of force of the magnetic field around the planet Jupiter, for example, give off synchrotron radiation at radio wavelengths. Synchrotron radiation at such wavelengths and at those of visible and ultraviolet light is generated by electrons moving in the magnetic field associated with the supernova remnant known as the Crab Nebula. Radio emissions of the synchrotron variety also have been detected from other supernova remnants in the Milky Way Galaxy and from extragalactic objects called quasars.

(Excerpt end)

Observation:

There are many X-ray point sources in the universe including one at the core of most spiral galaxies. These sources were described in detail in the author's book Cosmology Transition.

As somewhat described in the excerpt above, all those X-ray sources have an electrical current whose path is bent by a magnetic field resulting in this broad spectrum of wave lengths spanning from X-ray to infrared.

Quasars are typically dimmed in the optical wave lengths by their surrounding clouds of gas and dust.

11.6 Thermal Radiation

Excerpt from Wikipedia:

Thermal radiation is electromagnetic radiation generated by the thermal motion of particles in matter. All matter with a temperature greater than absolute zero emits thermal radiation.

If a radiation object meets the physical characteristics of a black body in thermodynamic equilibrium, the radiation is called blackbody radiation. Planck's law describes the spectrum of blackbody radiation, which depends solely on the object's temperature. Wien's displacement law determines the most likely frequency of the emitted radiation, and the Stefan–Boltzmann law gives the radiant intensity for the wave length.

(Excerpt end)

Thermal radiation is also one of the fundamental mechanisms of heat transfer. Conduction between adjacent solid objects is another.

Its spectrum is characterized by a wave length distribution, with the wave length having the highest intensity related to the object's temperature.

The wave length distribution affects whether it is visible. A cool temperature won't be. When warmer the increasing infrared intensity can be felt as heat or warmth but not seen. A rising temperature will become visible as red. When even hotter the mix of color wave lengths can result in "white hot." Our Sun is hot enough to generate the ultraviolet frequency which is not visible but can affect the eyes and skin.

Our white Sun can appear yellow when overhead due to the wave length distribution after the light passes through our atmosphere. The atmosphere can also cause a color change between sun rise and sun set, and it causes the sky to be blue.

Here is the thermal radiation spectrum from our Sun (from Wikipedia)

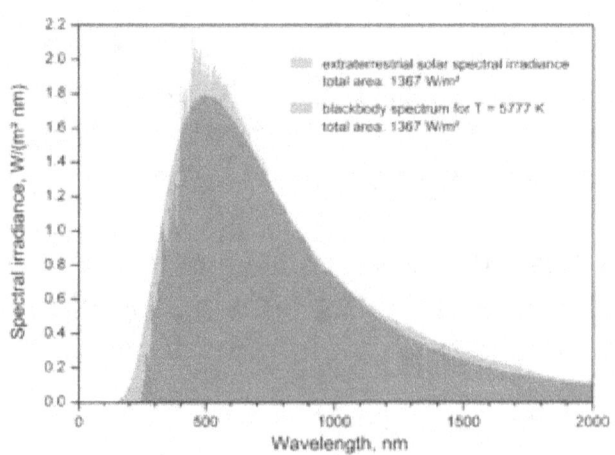

Important note about wave lengths:

Thermal radiation typically spans from ultraviolet to infrared to wave lengths covering most temperatures.

Infrared is always present but shorter wave lengths arise with a high enough surface temperature. Our Sun's thermal radiation, seen as light, is in this wave length range of UV to infrared.

Most emission lines from atoms range from visible to ultraviolet wave lengths. As a general rule, any wave lengths measured outside of this range, like radio at the low end, and X-ray or gamma ray at the high end, were emitted by a source of synchrotron radiation.

A fictitious black hole violates this general rule because the impossible hot accretion disk is claimed to emit X-rays but that energy requires an impossible temperature.

Thermal radiation requires a surface.

The temperature of a gas is measured by the kinetic energy of its atoms or molecules. A gas cannot emit thermal radiation. When its atoms and molecules become ionized, then as each ion captures an electron, they emit their characteristic wave length of electromagnetic radiation. This is the non-thermal mechanism for the color of a neon light.

11.7 Spectrum wave lengths and ranges

For a convenient reference, here are various wave lengths and their ranges.

gamma rays 1pm or E-12 m
hard x-rays 10pm or E-11 m
soft x-rays 100 pm or E-10 m or Angstroms
extreme UV 10 nm or E-9 m
near UV 100 nm or E-7 m
UVC 100-280 nm
UVB 280-315 nm
UVA 315-400 nm

visible colors are between 380-750 nm
The visible colors in Angstroms:

Violet 4000
Blue 4600
Cyan 4900
Green 5000
Yellow 5800
Orange 6000
Red 7000

near infrared 1 um or E-6 m
mid infrared 10 um or E-5 m
far infrared 100 um or E-4 m
EHF 1 mm or E-3 m
SHF 10 mm or E-2 m
UHF 100 mm or E-1 m
VHF 1 m
HF 10 m

11.7 Hierarchy of Reactions to light

An atom in the path of electromagnetic radiation must do one of the following:
1) Particle pair production,
2) Ionization,
3) Compton scattering,
4) Photoelectric effect,
5) Absorption line,
6) Reflect or re-emit it,
7) Transfer to vibration in a molecular bond,
8) Transfer to kinetic energy of the atom or molecule.

If (1) can be done, then the action is performed.
If not, the possible sequence of actions continues.

If (2) can be done, then the action is performed.
If not, the sequence continues.
If (3) can be done, then the action is performed.
If not, the sequence continues.
If (4) can be done, then the action is performed.
If (5) can be done, then the action is performed.
If (6) can be done, then the action is performed.
If (7) can be done, then the action is performed.
Action (7 or 8) must be done, if no other.

Energy must be transferred, as in (1,2,3.4,5,6,7), or transformed, as in (8).

Actions (1,2,5) absorb only some of the incoming energy but the total energy is conserved through the partial transfer.

Action (1) has two requirements. First, the atom's outer shell must have enough electrons because 2 will be ejected.

Descriptions of this action do not list the elements tested for this action.

The second requirement is the atom's state of matter. The description of its observations mention a "cloud chamber" implying this action has been observed only with unbound atoms in a gas.

Because the descriptions of this action lack all the necessary details, it is impossible to thoroughly explain its requirements. The mechanism can be explained. That explanation is in section 6.

Action (2) can occur in any state. If the radiation in its continuum has the energy for the outer shell to eject an electron, then in that instant, the energy is transferred to the ejected electron's kinetic energy.

Electromagnetic radiation continues its propagation until either absorbed or transferred.

Action (3) is described separately below. This is a behavior on a surface.

Action (4) is described separately in section 5.

Action (5) occurs in an atom or molecule in a gas.

The atom can be neutral or ionized, but must have at least one electron. Doppler effect was explained earlier in this section 11.

Actions (3,4,6,7) can occur in a liquid or solid surface.

Action (8) is an instantaneous transfer of energy from the wave length to the particle's kinetic energy.

Thermal energy in a gas is held in the kinetic energy of its particles. Pressure and volume also affect the temperature of a gas.

11.8 Compton Scattering

Compton scattering is an atomic behavior involving the absorption of energy at a level between that required for particle pair production and the photoelectric effect.

Excerpt from Wikipedia:

Compton scattering, discovered by Arthur Holly Compton, is the scattering of a photon by a charged particle, usually an electron. If it results in a decrease in energy (increase in wavelength) of the photon (which may be an X-ray or gamma ray photon), it is called the Compton effect. Part of the energy of the photon is transferred to the recoiling electron. Inverse Compton scattering occurs when a charged particle transfers part of its energy to a photon.

Compton found that some X-rays experienced no wavelength shift despite being scattered through large angles; in each of these cases the photon failed to eject an electron. Thus the magnitude of the shift is related not to the Compton wavelength of the electron, but to the Compton wavelength of the entire atom, which can be upwards of 10000 times smaller. This is known as "coherent" scattering off the entire atom since the atom remains intact, gaining no internal excitation.

In Compton's original experiments the wavelength shift given above was the directly-measurable observable. In modern experiments it is conventional to measure the energies, not the wavelengths, of the scattered photons.

(Excerpt end)

Observation:

This behavior is assumed to be an interaction between a photon and an electron, and claimed to be confirmation of light as a particle, the photon, and not a wave length.

This is an outright contradiction to Compton's conclusion the emitted wave length is determined by the "entire atom."

There is no photon interacting with a free charged particle. An atom's outer shell electrons are absorbing the energy required to change its energy to one which is exactly acceptable as defined by the atom and its electrons.

This behavior is the intermediate result between the other energy levels for the atom. Photoelectric effect results in an electron ejection, with the excess energy transferred to the departing electron.

The explanation for Compton scattering requires the atom absorb the necessary energy, for this action, among its electrons thereby increasing its energy level held among them. That energy must be released when the atom returns to its ground state.

Unlike the photoelectric effect having 1 action, at this higher energy level of Compton scattering, the atom performs 2 electron actions.

1) 1 electron is ejected.

2) A second electron moved to ground state resulting in the radiation for that change in its energy. This charge moved a very short distance resulting in a very short emission line.

The description states this wave length change is not quantized.

Excerpt from above:

"The wavelength shift is at least zero and at most twice the Compton wavelength of the electron."

The Wikipedia image and description of Compton's experiment in 1923 shows a "graphite target" suggesting a target's surface having a lattice of carbon Carbon has 6 electrons. There is no mention of a list of the other elements having this behavior measured with consistent results.

By comparison, the higher energy particle pair production (PPairP) also affects 2 electrons in the atom.
The difference between them is PPairP ejects the second electron as a positron, while CS gets a wave length, increased than that absorbed by the atom, from the non-ejected electron.

Compton scattering (CS) is a wave length not a photon behavior.

11.9 Reflection

After light is emitted, it continues its propagation until absorbed.

Some surfaces, having the lattice structure of condensed matter can absorb and re-emit the incoming energy.

This action is observed with the surface of water or glass.

A mirror has the reflective surface behind the transparent glass.

The transparent glass has a diffraction index so the velocity of ligh's propagation slows through the glass without being absorbed.

After passing through the glass, the propagation is affected by the current medium. When the back surface is not
transparent, it must absorb or re-emit the light.

11.10 Molecular Vibration

Excerpt from Wikipedia:

A molecular vibration is a periodic motion of the atoms of a molecule relative to each other, such that the center of mass of the molecule remains unchanged.
The typical vibrational frequencies, range from less than 1013 Hz to approximately 1014 Hz, corresponding to wavenumbers of approximately 300 to 3000 cm−1.
In general, a non-linear molecule with N atoms has 3N − 6 normal modes of vibration, but a linear molecule has 3N − 5 modes, because rotation about the molecular axis cannot be observed. A diatomic molecule has one normal mode of vibration, since it can only stretch or compress the single bond. Vibrations of polyatomic molecules are described in terms of normal modes, which are independent of each other, but each normal mode involves simultaneous vibrations of different parts of the molecule.
A molecular vibration is excited when the molecule absorbs energy, ΔE, corresponding to the vibration's frequency, v, according to the relation ΔE = hv, where h is Planck's constant. A fundamental vibration is evoked when one such quantum of energy is absorbed by the molecule in its ground state. When multiple quanta are absorbed, the first and possibly higher overtones are excited.

(Excerpt end)

This book is not about chemistry. The scope of this book covers the atomic model and not all behaviors observed in the science of chemistry.

The point here is energy must be transferred or transformed but must be conserved in all events.

The excerpt contains Lori Gardi's change to Planck's equation.

12 Atomic Equilibrium

An atom exists in a state of equilibrium.

An atom can make changes in an instant. These changes can be only in the electrons, only in the nucleus, or in both.

Among these instantaneous changes:

Alpha decay,
Beta decay,
Gamma decay,
Electron capture.

Each item will be described in turn.

Before the individual behaviors, there is another critical behavior to note. There is a particle competition within the nucleus. A context in the nucleus determines its decay selection.

Excerpt from Wikipedia:

Usually unstable nuclides are clearly either "neutron rich" or "proton rich", with the former undergoing beta decay and the latter undergoing electron capture (or more rarely, due to the higher energy requirements, positron decay). However, in a few cases of odd-proton, odd-neutron radionuclides, it may be energetically favorable for the radionuclide to decay to an even-proton, even-neutron isobar either by undergoing beta-positive or beta-negative decay.

An often-cited example is the single isotope ^{64}Cu (29 protons, 35 neutrons), which illustrates three types of beta decay in competition.

Copper-64 has a half-life of about 12.7 hours. This isotope has one unpaired proton and one unpaired neutron, so either the proton or the neutron can decay. This particular nuclide (though not all nuclides in this situation) is almost equally likely to decay through proton decay by positron emission (18%) or electron capture (43%) to ^{64}Ni, as it is through neutron decay by electron emission (39%) to ^{64}Zn.

(Excerpt end)

Observation:

Whether a nucleus is proton-rich or neutron-rich determines which beta decay occurs. Both are described below.

12.1 Electron capture

The electron capture step of radioactive decay involves a proton in the nucleus capturing one of the electrons in orbit at some distance.

The capture results in a) the proton changing into a neutron, b) a drop in the nucleus positive charge occurring at the same instant with c) the drop in the electron count, d) the difference between the number of electrons and protons remains the same.

The electron capture step can be important when anyone is using certain isotopes.

From Wikipedia: "The isotope technetium-97 decays only by electron capture, and could be inhibited from radioactive decay by fully ionizing it."

The electron capture event indicates there is a type of equilibrium between a nucleus and its set of electrons. There 2 sets of participants, with a number of protons matching the number of electrons, if the atom was neutral, not ionized, at the instant of the capture.

The trigger for any radioactive step is unknown, as well as the cause of an observed half-life, rather than some other duration. This ^{97}Tc nucleus can wait a long time before it pulls in an electron. ^{97}Tc has a half-life of 4.21×10^6 y.

Several radioactive behaviors were identified above.

12.2 Alpha Decay

Alpha decay requires the nucleus has the structure where an alpha particle already exists on the periphery of the nucleus. At the moment of instability, this particle of 2 protons bound to 2 neutrons is ejected by the Coulomb's force between a positive alpha particle and the rest of the nucleus which is also positive. The force for this ejection is sometimes called the weak force. Alpha decay occurs in the heaviest elements, apparently starting at Tellurium isotopes which have 59 protons. Several of its isotopes having 52 or more neutrons do alpha decay.

12.3 Beta Decays

Beta decay requires the nucleus change its charge by one electron charge, by either an increase by emitting $1e^-$ (beta minus) or a decrease by emitting $1e^+$ (beta plus).

The two decays are explained further below.

Sometimes, beta decay has a gamma ray.

12.4 Gamma Decay

Gamma decay is not clearly described. This step is usually associated with the element radium. It is also often associated with the alpha decay.

Excerpt from Wikipedia:

A sample of radium metal maintains itself at a higher temperature than its surroundings because of the radiation it emits – alpha particles, beta particles, and gamma rays.

More specifically, natural radium (which is mostly ^{226}Ra) emits mostly alpha particles, but other steps in its decay chain (the uranium or radium series) emit alpha or beta particles, and almost all particle emissions are accompanied by gamma rays.

In 2013, it was discovered that the nucleus of radium-224 is pear-shaped. This was the first discovery of an asymmetric nucleus.

   (Excerpt end)

The radium excerpt offers insight into radioactive decay.

First, a nucleus lacking symmetry enables a loss in stability when equilibrium between forces is disturbed.

Second, a spectrum is never provided for the gamma ray detection.

Synchrotron radiation when the motion of charged particles is diverted by a magnetic field. The peak frequency is determined by the velocity of the particles. This the mechanism for generating X-rays like for medical imaging.

Alpha decay is a particle, having 2 protons and 2 neutrons, being ejected at high velocity, so this is charge in motion.

From Wikipedia about the alpha particle:

"Due to the mechanism of their production in standard alpha radioactive decay, alpha particles generally have a kinetic energy of about 5 MeV, and a velocity in the vicinity of 4% of the speed of light."

If the spectrum of radiation from radium were ever measured, it will be synchrotron radiation which is a relatively flat wave length distribution with infrared at the high end and by this observation the gamma ray wavelength is present in the mix at the low end. Radium is also considered to radiate heat which is expected with infrared included.

The alpha particles apparently have the great velocity required for the energy of gamma rays to be propagated by their ejection from the nucleus.

12.5 Beta Plus Decay

There are 2 beta decays, beta-plus or beta-minus.

Beta plus decay is also called positron emission.

Excerpt from Wikipedia:

Positron emission or beta plus decay (β+ decay) is a subtype of radioactive decay called beta decay, in which a proton inside a radionuclide nucleus is converted into a neutron while releasing a positron and an electron neutrino (ve). Positron emission is mediated by the weak force. The positron is a type of beta particle (β+), the other beta particle being the electron (β−) emitted from the β− decay of a nucleus.

An example of positron emission (β+ decay) is magnesium-23 decaying into sodium-23.

Because positron emission decreases proton number relative to neutron number, positron decay happens typically in large "proton-rich" radionuclides. Positron decay results in nuclear transmutation, changing an atom of one chemical element into an atom of an element with an atomic number that is less by one unit.

A positron is ejected from the parent nucleus, and the daughter (Z−1) atom must shed an orbital electron to balance charge. The overall result is that the mass of two electrons is ejected from the atom (one for the positron and one for the electron).

(Excerpt end)

Observation:

This behavior is like electron capture.

Both are protons -1 and neutrons +1

The only apparent difference is the positron emission, but with a missing electron.

I interpret this description as a neutron is created while a positron is ejected and an electron is "shed" which probably means missing so the final atom is an ion.

This is the scenario with ^{23}Mg and ^{23}Na

Start :
^{23}Mg has 12p + 11n, 12 e in orbit

End:
^{23}Na has 11p + 12n, 10 e in orbit because 1 electron was "shed" while 1 positron was ejected also

The instantaneous event when everything is simultaneous:
a) 1e was captured from K shell by a proton to become a neutron.
b) This instant has the 12p in nucleus +1 compared to remaining 11 electrons in orbit.

c) This captured electron's polarity is flipped from normal e– to e+ becoming a positron.

d) Ejecting that "flipped in an instant" positron at +1 balances the charges.

e) Therefore, the positron is the captured electron but its polarity was flipped. The positron did not come from a proton.
f) Another, second electron from K shell is captured.

g) This instant has the nucleus at 11p + 12p again
The stated changes are from 12 to 10 electrons in orbit, so one electron is captured to form a neutron, while one positron is ejected for mass balance.

At the moment an electron flips its polarity while adjacent to a proton, the two + charges repel, causing the positron ejection.

^{23}Na is known to decay with β+ emission having a half-life of 22.422 seconds.

Positron creation was also described in section 6 Particle Pair Production.

12.6 Beta-minus decay

Above, with nucleus competition description, beta minus decay occurs in a "neutron rich" nucleus. As a result, one neutron ejects its accompanying electron due to too many electrons among the neutrons. Equilibrium among these protons and their electrons is restored by ejecting one of the electrons.

Excerpt from Wikipedia:

> In nuclear physics, beta decay (β-decay) is a type of radioactive decay in which a beta particle (fast energetic electron or positron) is emitted from an atomic nucleus, transforming the original nuclide to an isobar. For example, beta decay of a neutron transforms it into a proton by the emission of an electron accompanied by an antineutrino.

Observation: Beta minus decay is accompanied by an antineutrino.

12.7 Neutrino

The Neutrino must be mentioned but it involves behaviors mentioned earlier in this book. There was no earlier, appropriate place for this particle's behaviors.

The neutrino is not fully understood so this topic requires some conjecture. For that reason, it is not given its own major section in the book. This book is about updating the atomic model to explain the mass defect behavior with an updated atomic model. A neutrino is never part of that particular behavior.

There are 3 types or flavors of a neutrino. Each is explained in turn.

12.7.1 Electron Neutrino

Excerpt from Wikipedia:

The electron neutrino (v_e) is a subatomic lepton elementary particle which has zero net electric charge.

   (Excerpt end)

Observation:

From the reference, the mass of this particle is "small but non-zero."

It is important to get the background for a neutrino and how its ambiguous mass value arose.

That is the Cowan–Reines neutrino experiment.

Excerpt from Wikipedia:

The Cowan–Reines neutrino experiment was conducted by Washington University in St. Louis alumnus Clyde L. Cowan and Stevens Institute of Technology and New York University alumnus Frederick Reines in 1956. The experiment confirmed the existence of neutrinos. Neutrinos, subatomic particles with no electric charge and very small mass, had been conjectured to be an essential particle in beta decay processes in the 1930s. With neither mass nor charge, such particles appeared to be impossible to detect. The experiment exploited a huge flux of (hypothetical) electron antineutrinos emanating from a nearby nuclear reactor and a detector consisting of large tanks of water. Neutrino interactions with the protons of the water were observed, verifying the existence and basic properties of this particle for the first time.

Only the resulting electron was observed, so its varying energy suggested that energy may not be conserved. This quandary and other factors led Wolfgang Pauli to attempt to resolve the issue by postulating the existence of the neutrino in 1930. If the fundamental principle of energy conservation was to be preserved, beta decay had to be a three-body, rather than a two-body, decay. Therefore, in addition to an electron, Pauli suggested that another particle was emitted from the atomic nucleus in beta decay. This particle, the neutrino, had very small mass and no electric charge; it was not observed, but it carried the missing energy.

(Excerpt end)

There is no loss of mass but the electron did not have the correct kinetic energy.

That difference in energy, from expected, varied. It was not "continuous spectrum" which is an odd term when there is no electromagnetic radiation here.

The solution was proposing a third body in the event, with the other 2 being proton and electron.

The third body cannot have mass because none can be created.

This third body eventually was named a neutrino.
This neutrino is inconsistent with particle physics because a elementary particle must have mass.

Both the electron and proton have mass and a defined size. The proton can be compressed in a nucleus so it must have a physical surface. A neutron is apparently the bond between proton and electron implying an electron is also a body with a surface. Inside each is an amount of energy. This energy is driving its reactivity to other masses. This reactivity to masses is affected by the proton's size. Inside each is another amount of energy which drives its reactivity to other charges. As noted elsewhere the polarity of this charge behavior can be flipped during certain events. If a particle has the charge behavior, it always has the same reactivity, like that of an electron.

Therefore a neutrino is somehow a discrete entity able of carrying a variable amount of energy. It has no mass or charge behavior.

To confirm the existence of a no-mass particle it must be directly detectable, by a method having no other explanation than this neutrino.

That is why neutrino detectors were designed and built. The Sudbury Neutrino Observatory is one.

Excerpt from Wikipedia:

The Sudbury Neutrino Observatory (SNO) was a neutrino observatory located 2100 m underground in Vale's Creighton Mine in Sudbury, Ontario, Canada. The detector was designed to detect solar neutrinos through their interactions with a large tank of heavy water.
The detector was turned on in May 1999,

In the charged current interaction, a neutrino converts the neutron in a deuteron to a proton. The neutrino is absorbed in the reaction and an electron is produced. Solar neutrinos have energies smaller than the mass of muons and tau leptons, so only electron neutrinos can participate in this reaction. The emitted electron carries off most of the neutrino's energy, on the order of 5–15 MeV, and is detectable. The proton which is produced does not have enough energy to be detected easily. The electrons produced in this reaction are emitted in all directions, but there is a slight tendency for them to point back in the direction from which the neutrino came.

In the neutral current interaction, a neutrino dissociates the deuteron, breaking it into its constituent neutron and proton. The neutrino continues on with slightly less energy, and all three neutrino flavours are equally likely to participate in this interaction. Heavy water has a small cross section for neutrons, but when neutrons are captured by a deuterium nucleus, a gamma ray (photon) with roughly 6 MeV of energy is produced. The direction of the gamma ray is completely uncorrelated with the direction of the neutrino.

Some of the neutrons produced from the dissociated deuterons make their way through the acrylic vessel into the light water jacket surrounding the heavy water, and since light water has a very large cross section for neutron capture, these neutrons are captured very quickly. Gamma rays of roughly 2.2 MeV are produced in this reaction, but because the energy of the photons is less than the detector's energy threshold (meaning they do not trigger the photomultipliers), they are not directly observable. However, when the gamma ray collides with an electron via Compton scattering, the accelerated electron can be detected through Cherenkov radiation.

(Excerpt end)

Observation:

The excerpt has "when neutrons are captured by a deuterium.." but I believe that phrase is a mistake and should be "when neutrinos are captured by a deuterium..."

The gamma ray detected must have a mechanism like the alpha particle ejection, described above. The electron velocity must be similar to the alpha particle's velocity to propagate a similar gamma ray wave length, though the experiment reports no measurement of this radiation's spectrum.

SNO does not perform a direct detection of a neutrino. It uses an indirect method by monitoring and looking at events having no visible cause.

The technique to detect an electron neutrino involves the detection of an ejected pair of electron and neutron along with a gamma ray emission from a pool of water containing deuterium, so both ^2H and ^1H are with the single oxygen atom. The proton "does not have enough energy to be detected easily." The more likely explanation is the proton, having lost its neutron, must remains with its orbiting electron, so the proton also must remain bound to the oxygen atom in this water molecule.

The assumption for this experiment is only the energy carried within an electron neutrino can break the bond between proton and neutron in the deuterium nucleus.

The excerpt continues with the results:

The first scientific results of SNO were published on 18 June 2001, and presented the first clear evidence that neutrinos oscillate (i.e. that they can transmute into one another), as they travel from the Sun. This oscillation, in turn, implies that neutrinos have non-zero masses. The total flux of all neutrino flavours measured by SNO agrees well with theoretical predictions. Further measurements carried out by SNO have since confirmed and improved the precision of the original result.

    (Excerpt end)

Observation:

This author disagrees with these conclusions. There was nothing observed during this experiment to justify them. This is an indirect measurement by looking for results and assuming no other causes are present to explain those results.

If the neutrino is claimed to have a non-zero mass then it must be measurable. For example, the Moon can pass between the Sun and the SNO. The Moon has a substantial non-zero mass so the neutrino must interact with the Moon by at least changing the neutrino's path. Therefore SNO should see detections affected by the Moon at that time. The Moon's location is predictable.

Certainly, there is no justification of a confirmed change in flavour, unless one knows with certainty its flavour at its source. It is impossible to measure a neutrino at its point of origin.

This experiment claims to detect the results of an encounter with something having no measurable attributes, other than an amount of energy it carries. For this book, this is not sufficient to consider as evidence.

12.7.2 Tau Neutrino

Excerpt from Wikipedia:

The tau neutrino or tauon neutrino is a subatomic elementary particle which has the symbol v_τ and no net electric charge. Together with the tau (τ), it forms the third generation of leptons, hence the name tau neutrino. Its existence was immediately implied after the tau particle was detected in a series of experiments between 1974–1977 by Martin Lewis Perl with his colleagues at the SLAC–LBL group. The discovery of the tau neutrino was announced in July 2000 by the DONUT collaboration (Direct Observation of the Nu Tau).

(Excerpt end)

Observation:

The tau particle exists only in particle accelerators. This book ignores whatever fragments or claimed particles are proposed by those using them. That rule was justified in the initial section Fundamental Particles.

Neutrinos are claimed to have 3 flavors but none have a measured mass. A neutrino needs more experimental evidence to draw conclusions on its currently inconsistent behaviors. It can be only a possible fundamental particle.

12.7.3 Antimatter Annihilation

A positron is ejected from several actions noted above.

One is created infrequently by an electron flipping its charge polarity. It is usually destroyed in a short time when it encounter an electron anywhere during is travel.

Excerpt from Wikipedia:

In particle physics, annihilation is the process that occurs when a subatomic particle collides with its respective antiparticle to produce other particles, such as an electron colliding with a positron to produce two photons. The total energy and momentum of the initial pair are conserved in the process and distributed among a set of other particles in the final state. Antiparticles have exactly opposite additive quantum numbers from particles, so the sums of all quantum numbers of such an original pair are zero.

Hence, any set of particles may be produced whose total quantum numbers are also zero as long as conservation of energy and conservation of momentum are obeyed. During a low-energy annihilation, photon production is favored, since these particles have no mass.

(Excerpt end)

Observation:

A positron is a disturbed particle having its charge polarity flipped. When the positron encounters an electron, their opposite charges will result in their collision. By observation the energy measured in each particle is released in radiation when the opposing internal energies meet This is because the positron is disturbed and cannot survive the encounter with its opposite so both do not survive the instant of the attempted merger. At the instant of merging each charge is moving a very short distance. As described in Section 4 Planck's Equation, energy is carried in the amplitude of a wave length. Each particle in motion initiates the propagation of a very short wave length, in the gamma ray range, based on the tiny distance. The amplitude of the wave length carries the energy which was carried in the particle to sustain its mass and charge behaviors. The 2 particles are approaching each other from opposing directions. The 2 opposing gamma ray wave length emissions are in the corresponding opposite directions.

13 Mass Defect Introduction

The phenomenon called mass defect requires a new explanation in particle physics.

It is not really a defect. It is just an observation of a different mass than expected.

On the atomic scale, a difference in mass can be observed between expected and measured in an atomic nucleus. The difference could be more or less than expected. The difference is called a mass defect. When it is less then expected, it can be called a mass deficit. The measured mass value of a particular isotope is never more than expected.

When someone is not careful and uses a mass value from the wrong isotope, then it is possible for that mass value to be greater than expected. This possibility was described in the author's book Redefining Gravity.

In this book, data for the isotopes of the elements are from Wikipedia. Links are in References.

The nucleus consists of protons and neutrons, but each neutron is a combination of proton and electron.

The mass defect is an apparent change in the protons and neutrons in the nucleus during the process of fusion requiring compression.

After this phenomenon has an explanation, then it should not be called a defect.

There are 2 sources to compare their descriptions of the behavior.

Mass defect has this description, from Britannica:

The observed atomic mass is slightly less than the sum of the masses of the protons, neutrons, and electrons that make up the atom. The difference, called the mass defect, is accounted for during the combination of these particles by conversion into binding energy, according to an equation in which the energy (E) released equals the product of the mass (m) consumed and the square of the velocity of light in vacuum (c); thus, $E = mc^2$.

(Excerpt end)

Observation:

A difference is not always explained by this mass/energy relationship. This book offers another explanation.

Wikipedia uses another name and has no topic for Mass Defect.

In Wikipedia, the topic "mass defect" refers to an anomaly in a spiral galaxy brightness profile near its core.

Wikipedia calls the atomic mass defect behavior something else.

Excerpt from Wikipedia:

Nuclear binding energy is the minimum energy that would be required to disassemble the nucleus of an atom into its component parts. These component parts are neutrons and protons, which are collectively called nucleons. The binding energy is always a positive number, as we need to spend energy in moving these nucleons, attracted to each other by the strong nuclear force, away from each other. The mass of an atomic nucleus is less than the sum of the individual masses of the free constituent protons and neutrons, according to Einstein's equation $E=mc^2$. This 'missing mass' is known as the mass defect, and represents the energy that was released when the nucleus was formed.

Mass defect (also called "mass deficit") is the difference between the mass of an object and the sum of the masses of its constituent particles. Discovered by Albert Einstein in 1905, it can be explained using his formula $E = mc^2$, which describes the equivalence of energy and mass. The decrease in mass is equal to the energy given off in the reaction of an atom's creation divided by c^2. By this formula, adding energy also increases mass (both weight and inertia), whereas removing energy decreases mass. For example, a helium atom containing four nucleons has a mass about 0.8% less than the total mass of four hydrogen nuclei (which contain one nucleon each). The helium nucleus has four nucleons bound together, and the binding energy which holds them together is, in effect, the missing 0.8% of mass.

(Excerpt end)

The entire periodic table is reviewed in the Periodic Table section to find whether any isotopes have "extra mass."

If this book's conclusions are accepted, the descriptions for mass defect must be updated, including the removal of mentions of Einstein and mass/energy calculations.

14 Isotope Data File

All the long life isotopes in the periodic table are analyzed for a mass defect

Here, a long life isotope is one which is either stable or its half-life is long enough for its measured mass value to have enough digits after the decimal for valid calculations among each set.
The author compiled data from all elements and their long life isotopes to compare each for their measured value against the sum of their components.
A reference file in the .zip format is available with this spreadsheet of element isotope data

ZIsotopes.zip

Note: The main worksheet, using MS Excel, has over 850 rows. Compressing that content into a smaller page is quite impractical.

The phenomenon called mass defect requires a new explanation in particle physics.

The process toward that goal begins with the analysis of this behavior among all the elements.

A work sheet with formulae expedites this analysis.

Each element has a mix of entries and calculations.

There are only 2 manual entries:

a) its defined number of protons,

b) the isotope's nominal atomic weight or sum of its nucleons,

This is entered twice; once as an integer; again as part of the element's isotope name.

There is 1 copy&paste entry:

Its measured atomic weight is saved as text, so there can be no changes, like from rounding. Consistency of digits after the decimal point enables the best comparison of values.

The masses of the electron and proton have 10 digits after the decimal point.

Microsoft Excel 2003 handles up to this precision. That is sufficient for this book.

From these few entries, the sum of an atom's particles can be calculated. This is called the predicted mass. These entries enable calculating a) the number of neutrons in the atom and b) the average mass of all the neutrons in the nucleus.

Unfortunately, few elements have their measured mass value stated with 10 digits after the decimal.

Therefore, those calculations inherently lack consistent accuracy with the particles.

This book is about explaining behaviors. The public data prevent consistent precision.

Consistent formulae in the spreadsheet perform identical steps of analysis for every isotope.

The measured mass value subtracts the predicted mass to obtain the mass defect for this element. It is the nature of an atomic nucleus to have a deficit due to the nucleon compression during fusion. When the measured is greater than predicted, there is an entry error.

EMI = Expected Mass of Isotope, or its Nominal Mass.

NP = number Protons, or the atomic number.

NN = number Neutrons, from NM - NP

MCN = Mass change per Neutron, from a calculation

The nominal mass of a neutron = $m_p + m_e$

EMI = (NP * m_p) + (NN * ($m_p + m_e$))

The simpler calculation of the isotope's expected mass:

EMI = NM * (mp + me)

The calculation of MCN:

MCN = (MM − EMI) / NN

In simple terms, each atom has a difference in mass between measured and expected, or predicted by the sum of its particles.

If the neutrons in the nucleus changed their apparent mass by MCN, then that explains that mass defect.

Every atom, except for the hydrogen isotope ^{1}H and the helium isotope ^{3}He, has more neutrons than protons in its nucleus.

The mass defect is more likely to be attributed to the neutrons, not the protons. The simplest way for an atom to become another isotope is by fusing another neutron into the nucleus. This book treats mass defect as a neutron behavior though it is actually a nucleus behavior. When using isotopes which are changes in the number of neutrons, there is little visibility to proton changes.

14.1 Calculating Neutron Mass

With a measured mass of a particular isotope, it is possible to calculate the mass of the neutrons by subtracting the mass of the protons and the orbiting electrons.

NM = Nominal Mass of the Isotope, or sum of its nucleons.

MM = Measured Mass of Isotope.

NP = number Protons, or the atomic number.

NN = number Neutrons, from NM – NP

The measured nominal mass of an atom is assumed to be this sum:

$$MM = (NP * (m_p + m_e)) + (NN * n^0)$$

Where m_e is the orbiting electron and n^0 is the neutron in the nucleus.

The average neutron mass can be calculated from:

$$n^0 = (MM - (NP * (m_p + m_e))) / NN$$

Average Neutron Mass will be used frequently so it will be ANM in the Section 15 Periodic Table.

The accepted mass of a neutron is never used for calculating the mass of an atom from its components because the accuracy of the neutron's measured mass is uncertain. The next section provides measurements of a neutron for ever element.

A neutron has no electrical charge and decays in a few minutes when outside a nucleus.

Therefore, its mass is currently calculated by analysis of deuterium whose nucleus is only 1 proton and 1 neutron.

Unfortunately, as noted elsewhere in this book, the process of creating a nucleus requires much force of compression so the result is a mass defect.

The author's analysis presented in section 12's data file calculates the neutron's mass for each consecutive isotope, where the difference is only 1 neutron between the 2 isotopes.

Because of the inevitable mass defect caused by variations in a fusion sequence, the calculated neutron mass from each isotope varies across the many isotopes, though the neutrons for one element's isotopes are similar within the set.

15 Periodic Table

All the long life isotopes in the periodic table are analyzed for a mass defect

Here, a long life isotope is one which is either stable or its half-life is long enough for its measured mass value to have enough digits after the decimal for valid calculations among each set.
The author compiled data from all elements and their long life isotopes to compare each for their measured value against the sum of their components.
Note: The main worksheet, using MS Excel, has over 850 rows. Compressing that content into a smaller page is quite impractical.

Since everyone does not have spreadsheet software, small portions of the spreadsheet, from each element, will be included in the book. For a complete picture of the mass defect behaviors across the periodic table, the spreadsheet should be consulted. The file information is in section 14 Isotope Data File.

Section 15 follows the numbering pattern of 15.X where X is the number of protons or the atomic number.

Each element will have these data when available:

ANM or Average Neutron Mass. The summed mass of the protons and electrons is subtracted from the isotope's measured mass and then divided by the number of neutrons to obtain this average mass per neutron. A nucleon (proton or neutron) cannot be measured individually.

A nominal neutron is 1 proton + 1 electron or 1.0078250322

Isotope	Change w/+Neutron	Unexpected Change
1H	0.0000000000	0.0000000000
2H	1.0062767459	-0.0015482864
3H	1.0019475039	-0.0058775284

These are the 3 columns:

1) Isotope has the atomic weight and the element
The isotope in bold is the element's nominal atomic weight.

2) Change w/+Neutron has the change in mass after adding 1 neutron. This value is from subtracting one isotope from the other. This change value is blank or zero for the first.

3) Unexpected Change has the difference between that change and a nominal neutron whose mass was shown above and is greater than 1.0

Among the many isotopes, the Unexpected Change value will vary from more or less than 1.0

15.1 Element 1 is Hydrogen or H

Hydrogen is the combination of 1 proton and 1 electron.

These are the hydrogen isotopes.

Isotope	Change w/+Neutron	Unexpected Change
H	0.0000000000	0.0000000000
2H	1.0062767459	-0.0015482864
3H	1.0019475039	-0.0058775284

The first neutron fused to the proton lost mass due to its compression. The second lost more than the first.

15.2.4 Element 2 is Helium or He

Helium has and 2 protons and 2 neutrons.

ANM = 0.9934765948

Isotope	Change w/+Neutron	Unexpected Change
3He		
4He	0.9865739315	-0.0212511008

Adding 1 neutron increased the atom's mass by less than 1.0

The average mass of the 2 neutrons (ANM) is less than 1.0

15.3 Element 3 is Lithium or Li

Lithium has and 3 protons and 4 neutrons.

ANM = 0.9981320851

6Li		0.0000000000
7Li	1.0008805496	-0.0069444826

15.4 Element 4 is Beryllium or Be

Beryllium has 4 protons and 5 neutrons.

ANM = 0.9961765882

9Be		0.0000000000
10Be	1.0013516300	-0.0064734022

15.5 Element 5 is Boron or B

Boron has 5 protons and 6 neutrons.

ANM = 0.9950300010

10B		0.0000000000
11B	0.9963683050	-0.0114567272

15.6 Element 6 is Carbon or C

Carbon has 6 protons and 6 neutrons.

Carbon-12 does not have a true measured mass. It has a value of "exactly 12" because it was used as the benchmark atom. Section 7 addressed this mistake.

ANM = 0.9921749678

12C		0.0000000000
13C		
14C	0.9998871528	-0.0079378795
15C	1.0073573120	-0.0004677202

15.7 Element 7 is Nitrogen or N

Nitrogen has 7 protons and 7 neutrons.

ANM = 0.9926141113

13N		0.0000000000
14N	0.9973353945	-0.0104896378
15N	0.9970348944	-0.0107901378
16N	1.0059930011	-0.0018320311

15.8 Element 8 is Oxygen or O

Carbon has 8 protons and 8 neutrons.

ANM = 0.9915392952

14O		0.0000000000
15O	0.9944688940	-0.0133561382
16O	0.9918490196	-0.0159760126
17O	1.0042171370	-0.0036078952
18O	1.0000278562	-0.0077971760

15.9 Element 9 is Fluorine or F

Fluorine has 9 protons and 10 neutrons.

ANM = 0.9927977873

17F		0.0000000000
18F	0.9988420600	-0.0089829722
19F	0.9974658629	-0.0103591693
20F	1.0015780871	-0.0062469451
21F	0.9999676500	-0.0078573822
22F	1.0030501000	-0.0047749322
23F	1.0005310000	-0.0072940322

15.10 Element 10 is Neon.

Neon has 10 protons and 10 neutrons.

ANM = 0.9914189854

18Ne		0.0000000000
19Ne	0.9961722000	-0.0116528322
20Ne	0.9905592762	-0.0172657560
21Ne	1.0014065138	-0.0064185184
22Ne	0.9975384200	-0.0102866122
23Ne	1.0030817900	-0.0047432422
24Ne	0.9991437000	-0.0086813322

15.11 Element 11 is Sodium, or Na

Sodium has 11 protons and 12 neutrons.

ANM = 0.9923373821

22Na		0.0000000000
23Na	0.9945532400	-0.0132717922
24Na	0.9909177570	-0.0169072752
25Na	1.0007952630	-0.0070297692
26Na	0.9967560100	-0.0110690222
27Na	1.0017476600	-0.0060773722
28Na	0.9995359700	-0.0082890622
29Na	1.0047404000	-0.0030846322

15.12 Element 12 is Magnesium or Mg

Magnesium has 12 protons and 12 neutrons.

ANM = 0.9909284425

22Mg		0.0000000000
23Mg	0.9945532400	-0.0132717922
24Mg	0.9909177570	-0.0169072752
25Mg	1.0007952630	-0.0070297692
26Mg	0.9967560100	-0.0110690222
27Mg	1.0017476600	-0.0060773722
28Mg	0.9995359700	-0.0082890622
29Mg	1.0047404000	-0.0030846322

15.13 Element 13 is Aluminum or Al

Aluminum has 13 protons and 14 neutrons.

ANM = 0.9914152136

24Al		0.0000000000
25Al	0.9904807700	-0.0173442622
26Al	0.9964635500	-0.0113614822
27Al	0.9946465500	-0.0131784822
28Al	1.0003716800	-0.0074533522
29Al	0.9985431100	-0.0092819222
30Al	1.0025148000	-0.0053102322

15.14 Element 14 is Silicon or Si

Silicon has 14 protons and 14 neutrons.

ANM = 0.9905268631

26Si		0.0000000000
27Si	0.9943708900	-0.0134541422
28Si	0.9902218450	-0.0176031872
29Si	0.9995681303	-0.0082569019
30Si	0.9972754717	-0.0105495605
31Si	1.0015930530	-0.0062319792
32Si	0.9987883100	-0.0090367222
33Si	1.0038255000	-0.0039995322
34Si	1.0005980000	-0.0072270322

15.15 Element 15 is Phosphorus or P

Phosphorus has 15 protons and 16 neutrons.

ANM = 0.9910241572

29P		0.0000000000
30P	0.9965130900	-0.0113119422
31P	0.9954485086	-0.0123765236
32P	1.0001456414	-0.0076793908
33P	0.9978180600	-0.0100069722
34P	1.0019202000	-0.0059048322
35P	0.9996682000	-0.0081568322

15.16 Element 16 is Sulfur or S

Sulfur has 16 protons and 16 neutrons.

ANM = 0.9904294162

30S		0.0000000000
31S	0.9946502400	-0.0131747922
32S	0.9925141644	-0.0153108678
33S	0.9993877355	-0.0084372967
34S	0.9964081001	-0.0114169321
35S	1.0011653100	-0.0066597222
36S	0.9980483800	-0.0097766522
37S	1.0040448100	-0.0037802222

15.17 Element 17 is Chlorine or Cl

Chlorine has 17 protons and 18 neutrons.

ANM = 0.9908792857

33Cl		0.0000000000
34Cl	0.9963104900	-0.0115145422
35Cl	0.9950902000	-0.0127348322
36Cl	0.9994541300	-0.0083709022
37Cl	0.9975957600	-0.0102292722
38Cl	1.0021078400	-0.0057171922
39Cl	0.9999977800	-0.0078272522
40Cl	1.0024118000	-0.0054132322

15.18 Element 18 is Argon or Ar

Argon has 18 protons and 22 neutrons.

ANM = 0.9918878429

36Ar		0.0000000000
37Ar	0.9992312050	-0.0085938272
38Ar	0.9959557900	-0.0118692422
39Ar	1.0015809000	-0.0062441322
40Ar	0.9980701238	-0.0097549084
41Ar	1.0021174762	-0.0057075560
42Ar	0.9985454000	-0.0092796322
43Ar	1.0025900000	-0.0052350322
44Ar	0.9992878000	-0.0085372322
45Ar	1.0031159000	-0.0047091322
46Ar	0.9999977000	-0.0078273322

15.19 Element 19 is Potassium or K

Potassium has 19 protons and 20 neutrons.

ANM = 0.9912058361

38K	0.9957052300	-0.0121198022
39K	0.9946253670	-0.0131996652
40K	1.0002916830	-0.0075333492
41K	0.9978270880	-0.0099979442
42K	1.0005770520	-0.0072479802
43K	0.9983323900	-0.0094926422
44K	1.0008523000	-0.0069727322
45K	0.9991045000	-0.0087205322
46K	1.0012901000	-0.0065349322
47K	0.9996800000	-0.0081450322
48K	1.0036796000	-0.0041454322
49K	1.0028696000	-0.0049554322

15.20 Element 20 is Calcium or Ca

Calcium has 20 protons and 20 neutrons.

ANM = 0.9903045111

40Ca		0.0000000000
41Ca	0.9996870540	-0.0081379782
42Ca	0.9963399100	-0.0114851222
43Ca	1.0001486000	-0.0076764322
44Ca	0.9967150700	-0.0111099622
45Ca	1.0007048000	-0.0071202322
46Ca	0.9975017000	-0.0103233322
47Ca	1.0008534000	-0.0069716322
48Ca	0.9979815000	-0.0098435322
49Ca	1.0030999800	-0.0047250522

15.21 Element 21 is Scandium or Sc

Scandium has 21 protons and 24 neutrons.

ANM = 0.9913160926

45Sc		0.0000000000
46Sc	0.9992600000	-0.0085650322
47Sc	0.9972356000	-0.0105894322
48Sc	0.9998235000	-0.0080015322
49Sc	0.9977930000	-0.0100320322
50Sc	1.0021640000	-0.0056610322

15.22 Element 22 is Titanium or Ti

Titanium has 22 protons and 26 neutrons.

ANM = 0.9913767535

46Ti		0.0000000000
47Ti	0.9991315000	-0.0086935322
48Ti	0.9961832000	-0.0116418322
49Ti	0.9999237000	-0.0079013322
50Ti	0.9969212000	-0.0109038322
51Ti	1.0018238000	-0.0060012322
52Ti	1.0002820000	-0.0075430322

15.23 Element 23 is Vanadium or V

Vanadium has 23 protons and 28 neutrons.

ANM = 0.991570848

47V		0.0000000000
48V	0.9973448000	-0.0104802322
49V	0.9962624000	-0.0115626322
50V	0.9986424000	-0.0091826322
51V	**0.9968010000**	-0.0110240322
52V	1.0008160000	-0.0070090322

15.24 Element 24 is Chromium or Cr

Chromium has 24 protons and 28 neutrons.

ANM = 0.9911680974

49Cr		0.0000000000
50Cr	0.9947085000	-0.0131165322
51Cr	0.9987232000	-0.0091018322
52Cr	0.9957401000	-0.0120849322
53Cr	1.0001419000	-0.0076831322
54Cr	0.9982310000	-0.0095940322
55Cr	1.0019593000	-0.0058657322
56Cr	0.9998134000	-0.0080116322

15.25 Element 25 is Manganese or Mn

Manganese has 25 protons and 30 neutrons.

ANM = 0.9914139765

51Mn		0.0000000000
52Mn	0.9973547000	-0.0104703322
53Mn	0.9957246000	-0.0121004322
54Mn	0.9990688000	-0.0087562322
55Mn	0.9976862000	-0.0101388322
56Mn	1.0008598000	-0.0069652322
57Mn	0.9993805000	-0.0084445322

15.26 Element 26 is Iron or Fe

Iron has 26 protons and 30 neutrons.

ANM = 0.9910495154

53Fe		0.0000000000
54Fe	0.9943011000	-0.0135239322
55Fe	0.9986844000	-0.0091406322
56Fe	0.9966429000	-0.0111821322
57Fe	1.0004565000	-0.0073685322
58Fe	0.9978816000	-0.0099434322
59Fe	1.0016011000	-0.0062239322

15.27 Element 27 is Cobalt or Co

Cobalt has 27 protons and 32 neutrons.

ANM = 0.9913099728

54Co		0.0000000000
55Co	0.9935394000	-0.0142856322
56Co	0.9978403000	-0.0099847322
57Co	0.9964521000	-0.0113729322
58Co	0.9994614000	-0.0083636322
59Co	0.9974422000	-0.0103828322
60Co	1.0006221000	-0.0072029322
61Co	0.9986587000	-0.0091663322

15.28 Element 28 is Nickel or Ni
Nickel has 28 protons and 31 neutrons.

ANM = 0.9908143806

57Ni		0.0000000000
58Ni	0.9955494000	-0.0122756322
59Ni	0.9990038000	-0.0088212322
60Ni	0.9964397000	-0.0113853322
61Ni	1.0002696000	-0.0075554322
62Ni	0.9972891000	-0.0105359322
63Ni	1.0013243000	-0.0065007322
64Ni	0.9982966000	-0.0095284322
65Ni	1.0021183000	-0.0057067322
66Ni	0.9990550000	-0.0087700322

15.29 Element 29 is Copper or Cu

Copper has 29 protons and 35 neutrons.

ANM = 0.9915096647

60Cu		0.0000000000
61Cu	0.9960928000	-0.0117322322
62Cu	0.9991262000	-0.0086988322
63Cu	0.9970135000	-0.0108115322
64Cu	1.0001667000	-0.0076583322
65Cu	0.9980253000	-0.0097997322
66Cu	1.0010793000	-0.0067457322
67Cu	0.9988615000	-0.0089635322
68Cu	1.0018806000	-0.0059444322
69Cu	0.9998184000	-0.0080066322
70Cu	1.0029630000	-0.0048620322

15.30 Element 30 is Zinc or Zn

Zinc has 30 protons and 35 neutrons.

ANM = 0.9912711438

63Zn		0.0000000000
64Zn	0.9959306000	-0.0118944322
65Zn	1.0000988000	-0.0077262322
66Zn	0.9967924000	-0.0110326322
67Zn	1.0010939000	-0.0067311322
68Zn	0.9977169000	-0.0101081322
69Zn	1.0017061000	-0.0061189322
70Zn	0.9987690000	-0.0090560322

15.31 Element 31 is Gallium or Ga

Gallium has 31 protons and 39 neutrons.

ANM = 0.9916578316

65Ga		0.0000000000
66Ga	0.9988542000	-0.0089708322
67Ga	0.9966127000	-0.0112123322
68Ga	0.9997784000	-0.0080466322
69Ga	0.9975935000	-0.0102315322
70Ga	1.0004484000	-0.0073766322
71Ga	0.9986793000	-0.0091457322
72Ga	1.0016650000	-0.0061600322
73Ga	0.9988084000	-0.0090166322
74Ga	1.0017713000	-0.0060537322
75Ga	0.9995542000	-0.0082708322
76Ga	1.0023274000	-0.0054976322
77Ga	1.0003267000	-0.0074983322
78Ga	1.0024539000	-0.0053711322

15.32 Element 32 is Germanium or Ge

Germanium has 32 protons and 41 neutrons.

ANM = .9920258017

69Ge		0.0000000000
70Ge	0.9962829000	-0.0115421322
71Ge	1.0007036000	-0.0071214322
72Ge	0.9971248000	-0.0107002322
73Ge	1.0013831000	-0.0064419322
74Ge	0.9977189000	-0.0101061322
75Ge	1.0016811000	-0.0061439322
76Ge	0.9985437000	-0.0092813322
77Ge	1.0021460000	-0.0056790322

15.33 Element 33 is Arsenic or As

Arsenic has 33 protons and 42 neutrons.

ANM = 0.9919850104

74As		0.0000000000
75As	0.9976678000	-0.0101572322
76As	1.0007975000	-0.0070275322
77As	0.9982533000	-0.0095717322

15.34 Element 34 is Selenium or Se

Selenium has 34 protons and 45 neutrons.

ANM = 0.9922766223

74Se		0.0000000000
75Se	1.0000470000	-0.0077780322
76Se	0.9966902000	-0.0111348322
77Se	1.0007004000	-0.0071246322
78Se	0.9973951000	-0.0104299322
79Se	1.0011900000	-0.0066350322
80Se	0.9980222000	-0.0098028322
81Se	1.0014712000	-0.0063538322

15.35 Element 35 is Bromine or Br

Bromine has 35 protons and 45 neutrons.

ANM = 0.9921034038

77Br		0.0000000000
78Br	0.9997670000	-0.0080580322
79Br	0.9971911000	-0.0106339322
80Br	1.0001922000	-0.0076328322
81Br	0.9977613000	-0.0100637322
82Br	1.0005135000	-0.0073115322
83Br	0.9983759000	-0.0094491322
84Br	1.0012990000	-0.0065260322
85Br	0.9991290000	-0.0086960322
86Br	1.0031900000	-0.0046350322
87Br	1.0019130000	-0.0059120322

15.36 Element 36 is Krypton or Kr

Krypton has 36 protons and 48 neutrons.

ANM = 0.9922876217

77Kr		0.0000000000
78Kr	0.9956948	-0.0121302322
79Kr	0.9997172	-0.0081078322
80Kr	0.996297	-0.0115280322
81Kr	1.000213	-0.0076120322
82Kr	0.9968916	-0.0109334322
83Kr	1.0006524	-0.0071726322
84Kr	0.997371	-0.0104540322
85Kr	1.0010203	-0.0068047322
86Kr	0.99808343	-0.0097416022
87Kr	1.00274413	-0.0050809022

15.37 Element 37 is Rubidium ot Rb

Rubidium has 37 protons and 48 neutrons.

ANM = 0.9921304905

85Rb		0
86Rb	0.999377682	-0.0084473502
87Rb	0.998013107	-0.0098119252

15.38 Element 38 is Strontium or Sr

Strontium has 38 protons and 50 neutrons.

ANM = 0.9921652206

84Sr		0.0000000000
85Sr	0.999508	-0.0083170322
86Sr	0.996327731	-0.0114973013
87Sr	0.999616766	-0.0082082661
88Sr	0.99673476	-0.0110902721
89Sr	1.001838443	-0.0059865893

15.39 Element 39 is Yttrium or Y

Strontium has 39 protons and 50 neutrons.

ANM = 0.9920134409

85Y		0.0000000000
86Y	0.998453	-0.0093720322
87Y	0.9959897	-0.0118353322
88Y	0.9986254	-0.0091996322
89Y	0.9963472	-0.0114778322
90Y	1.0013036	-0.0065214322
91Y	1.0001531	-0.0076719322
92Y	1.001644	-0.0061810322
93Y	1.000634	-0.0071910322
94Y	1.002012	-0.0058130322
95Y	1.001226	-0.0065990322
96Y	1.00307	-0.0047550322
97Y	1.002243	-0.0055820322

15.40 Element 40 is Zirconium or Zr

Zirconium has 40 protons and 51 neutrons.

ANM = 0.9920126375

85Zr		0.0000000000
86Zr	0.995	-0.0128250322
87Zr	0.998346	-0.0094790322
88Zr	0.995411	-0.0124140322
89Zr	0.998663	-0.0091620322
90Zr	0.9958144	-0.0120106322
91Zr	1.0009414	-0.0068836322
92Zr	0.999395	-0.0084300322
93Zr	1.0014352	-0.0063898322
94Zr	0.9998392	-0.0079858322
95Zr	1.0017274	-0.0060976322
96Zr	1.0002308	-0.0075942322
97Zr	1.0026797	-0.0051453322

15.41 Element 41 is Niobium or Nb

Niobium has 41 protons and 52 neutrons.

ANM = 0.9920298419

87Nb		0.0000000000
88Nb	0.99797	-0.0098550322
89Nb	0.995088	-0.0127370322
90Nb	0.997847	-0.0099780322
91Nb	0.995731	-0.0120940322
92Nb	1.000198	-0.0076270322
93Nb	0.9991841	-0.0086409322
94Nb	1.0009058	-0.0069192322
95Nb	0.9995519	-0.0082731322
96Nb	1.0012652	-0.0065598322
97Nb	0.9999976	-0.0078274322

15.42 Element 42 is Molybdenum or Mo

Molybdenum has 42 protons and 54 neutrons.

ANM = 0.9921486694

92Mo		0
93Mo	1.000002	-0.0078230322
94Mo	0.9982753	-0.0095497322
95Mo	1.0007538	-0.0070712322
96Mo	0.9988374	-0.0089876322
97Mo	1.001342	-0.0064830322
98Mo	0.99938332	-0.0084417122
99Mo	1.00230708	-0.0055179522
100Mo	0.9997651	-0.0080599322

15.43 Element 43 is Technetium or Tc

Technetium has 43 protons and 55 neutrons.

ANM = 0.9921952657

97Tc		0.0000000000
98Tc	1.000851	-0.0069740322
99Tc	0.9990387	-0.0087863322
100Tc	1.0014031	-0.0064219322

15.44 Element 44 is Ruthenium or Ru

Ruthenium has 44 protons and 57 neutrons.

ANM = 0.9923031698

95Ru		0.0000000000
96Ru	0.997185	-0.0106400322
97Ru	0.999957	-0.0078680322
98Ru	0.997732	-0.0100930322
99Ru	1.0006523	-0.0071727322
100Ru	0.9982802	-0.0095448322
101Ru	1.0013626	-0.0064624322
102Ru	0.9987672	-0.0090578322
103Ru	1.0019745	-0.0058505322
104Ru	0.9991092	-0.0087158322
105Ru	1.00232	-0.0055050322
106Ru	0.999576	-0.0082490322
107Ru	1.002581	-0.0052440322
108Ru	1.00026	-0.0075650322
109Ru	1.00303	-0.0047950322
110Ru	1.00094	-0.0068850322
111Ru	1.00356	-0.0042650322
112Ru	1.00127	-0.0065550322

15.45 Element 45 is Rhodium or Rh

Rhodium has 45 protons and 58 neutrons.

ANM = 0.9922996129

102Rh		0.0000000000
103Rh	0.998661	-0.0091640322
104Rh	1.001152	-0.0066730322
105Rh	0.999038	-0.0087870322

15.46 Element 46 is Palladium or Pd

Palladium has 46 protons and 61 neutrons.

ANM = 0.9925439593

101Pd		0.0000000000
102Pd	0.99732	-0.0105050322
103Pd	1.000478	-0.0073470322
104Pd	0.997949	-0.0098760322
105Pd	1.001049	-0.0067760322
106Pd	0.998401	-0.0094240322
107Pd	1.001647	-0.0061780322
108Pd	0.998759	-0.0090660322
109Pd	1.002058	-0.0057670322
110Pd	0.999203	-0.0086220322
111Pd	1.002518	-0.0053070322
112Pd	0.999643	-0.0081820322
113Pd	1.002836	-0.0049890322
114Pd	1.000213	-0.0076120322
115Pd	1.003317	-0.0045080322
116Pd	1.00048	-0.0073450322
117Pd	1.00368	-0.0041450322
118Pd	1.00114	-0.0066850322

15.47 Element 47 is Silver or Ag

Silver has 47 protons and 61 neutrons.

ANM = 0.9924291719

104Ag		0.0000000000
105Ag	0.9979	-0.0099250322
106Ag	1.00014	-0.0076850322
107Ag	0.998428	-0.0093970322
108Ag	1.000859	-0.0069660322
109Ag	0.998796	-0.0090290322
110Ag	1.001355	-0.0064700322
111Ag	0.999184	-0.0086410322
112Ag	1.001714	-0.0061110322
113Ag	0.999562	-0.0082630322
114Ag	1.002237	-0.0055880322
115Ag	0.999956	-0.0078690322
116Ag	1.0026	-0.0052250322
117Ag	1.00032	-0.0075050322
118Ag	1.0029	-0.0049250322
119Ag	1.00109	-0.0067350322

15.48 Element 48 is Cadmium or Cd

Cadmium has 48 protons and 64 neutrons.

ANM = 0.9926118164

103Cd		0.0000000000
104Cd	0.99643	-0.0113950322
105Cd	0.999619	-0.0082060322
106Cd	0.996991	-0.0108340322
107Cd	1.000159	-0.0076660322
108Cd	0.997566	-0.0102590322
109Cd	1.000798	-0.0070270322
110Cd	0.9980201	-0.0098049322
111Cd	1.001176	-0.0066490322
112Cd	0.9985797	-0.0092453322
113Cd	1.0016439	-0.0061811322
114Cd	0.9989568	-0.0088682322
115Cd	1.0020725	-0.0057525322
116Cd	0.999325	-0.0085000322

15.49 Element 49 is Indium or In

Indium has 49 protons and 66 neutrons.

ANM = 0.9927341124

103In		0.0000000000
104In	0.998386	-0.0094390322
105In	0.996374	-0.0114510322
106In	0.998791	-0.0090340322
107In	0.99683	-0.0109950322
108In	0.999403	-0.0084220322
109In	0.997453	-0.0103720322
110In	1.000014	-0.0078110322
111In	0.997938	-0.0098870322
112In	1.000429	-0.0073960322
113In	0.998526	-0.0092990322
114In	1.000856	-0.0069690322
115In	0.998964	-0.0088610322
116In	1.001382	-0.0064430322
117In	0.999254	-0.0085710322
118In	1.00184	-0.0059850322
119In	0.999491	-0.0083340322

15.50 Element 50 is Tin or Sn

Tin has 50 protons and 69 neutrons.

ANM = 0.9929283534

		0
112Sn		0
113Sn	1.000353	-0.0074720322
114Sn	0.997608	-0.0102170322
115Sn	1.000563	-0.0072620322
116Sn	0.998399	-0.0094260322
117Sn	1.001211	-0.0066140322
118Sn	0.998651	-0.0091740322
119Sn	1.001705	-0.0061200322
120Sn	0.9988867	-0.0089383322
121Sn	1.0020408	-0.0057842322
122Sn	0.9992035	-0.0086215322
123Sn	1.0022818	-0.0055432322
124Sn	0.9995531	-0.0082719322
125Sn	1.0025102	-0.0053148322
126Sn	0.9998689	-0.0079561322
127Sn	1.002707	-0.0051180322
128Sn	1.000177	-0.0076480322
129Sn	1.002943	-0.0048820322
130Sn	1.000487	-0.0073380322
131Sn	1.003033	-0.0047920322
132Sn	1.000816	-0.0070090322

15.51 Element 51 is Antimony or Sb

Antimony has 51 protons and 71 neutrons.

ANM = 0.9928405124

115Sb		0.0000000000
116Sb	1.000196	-0.0076290322
117Sb	0.998042	-0.0097830322
118Sb	1.000693	-0.0071320322
119Sb	0.998413	-0.0094120322
120Sb	1.00113	-0.0066950322
121Sb	0.9987437	-0.0090813322
122Sb	1.001358	-0.0064670322
123Sb	0.9990403	-0.0087847322
124Sb	1.0017217	-0.0061033322
125Sb	0.9993181	-0.0085069322

15.52 Element 52 is Tellurium or Te

Tellurium has 52 protons and 76 neutrons.

ANM = 0.9933889661

120Te		0.0000000000
121Te	1.000916	-0.0069090322
122Te	0.9981079	-0.0097171322
123Te	1.0012261	-0.0065989322
124Te	0.9985479	-0.0092771322
125Te	1.0016128	-0.0062122322
126Te	0.998881	-0.0089440322
127Te	1.0019146	-0.0059104322
128Te	0.9992368	-0.0085882322
129Te	1.0021351	-0.0056899322
130Te	0.9996262	-0.0081988322
131Te	1.0022995	-0.0055255322

15.53 Element 53 is Iodine or I

Iodine has 53 protons and 74 neutrons.

ANM = 0.9931046796

125I			0.0000000000
126I		1.0009938	-0.0068312322
127I		0.998849	-0.0089760322
128I		1.001336	-0.0064890322
129I		0.999179	-0.0086460322
130I		1.001686	-0.0061390322
131I		0.9994506	-0.0083744322
132I		1.0018724	-0.0059526322
133I		0.9998	-0.0080250322
134I		1.001947	-0.0058780322

15.54 Element 54 is Xenon or Xe

Xenon has 54 protons and 77 neutrons.

ANM = 0.9932796189

126Xe		0
127Xe	1.00091	-0.0069150322
128Xe	0.9983473	-0.0094777322
129Xe	1.0012481	-0.0065769322
130Xe	0.9987286	-0.0090964322
131Xe	1.0015744	-0.0062506322
132Xe	0.9990711	-0.0087539322
133Xe	1.0017572	-0.0060678322
134Xe	0.9994838	-0.0083412322
135Xe	1.0018325	-0.0059925322
136Xe	0.999992	-0.0078330322
137Xe	1.004343	-0.0034820322

15.55 Element 55 is Caesium or Cs

Cesium has 55 protons and 78 neutrons.

ANM = 0.9932701944

131Cs		0.0000000000
132Cs	1.0009703	-0.0068547322
133Cs	0.999017633	-0.0088073992
134Cs	1.001266542	-0.0065584902
135Cs	0.999258525	-0.0085665072
136Cs	1.0013346	-0.0064904322

15.56 Element 56 is Barium or Ba

Barium has 56 protons and 81 neutrons.

ANM = 0.9934274765

129Ba		0.0000000000
130Ba	0.9976418	-0.0101832322
131Ba	1.0006202	-0.0072048322
132Ba	0.9981203	-0.0097047322
133Ba	1.0009462	-0.0068788322
134Ba	0.9985009	-0.0093241322
135Ba	1.0011802	-0.0066448322
136Ba	0.9988873	-0.0089377322
137Ba	1.0012515	-0.0065735322
138Ba	0.9994198	-0.0084052322
139Ba	1.0035941	-0.0042309322
140Ba	1.0017637	-0.0060613322
141Ba	1.003806	-0.0040190322
142Ba	1.002042	-0.0057830322

15.57 Element 57 is Lanthanum or La

Lanthanum has 57 protons and 82 neutrons.

ANM = 0.9934186154

134La		0.0000000000
135La	0.998463	-0.0093620322
136La	1.000663	-0.0071620322
137La	0.998854	-0.0089710322
138La	1.000618	-0.0072070322
139La	0.9992413	-0.0085837322
140La	1.0031243	-0.0047007322
141La	1.0014844	-0.0063406322
142La	1.003117	-0.0047080322
143La	1.001984	-0.0058410322

15.58 Element 58 is Cerium or Ce

Cerium has 58 protons and 82 neutrons.

ANM = 0.9933120345

132Ce		0.0000000000
133Ce	1.000055	-0.0077700322
134Ce	0.99741	-0.0104150322
135Ce	1.000226	-0.0075990322
136Ce	0.998021	-0.0098040322
137Ce	1.000634	-0.0071910322
138Ce	0.998185	-0.0096400322
139Ce	1.000662	-0.0071630322
140Ce	0.9987857	-0.0090393322
141Ce	1.0028376	-0.0049874322
142Ce	1.0009677	-0.0068573322
143Ce	1.003142	-0.0046830322
144Ce	1.001261	-0.0065640322

15.59 Element 59 is Praseodymium or Pr

Praseodymium has 59 protons and 82 neutrons.

ANM = 0.9932436085

141Pr		0.0000000000
142Pr	1.002392	-0.0054330322
143Pr	1.0007721	-0.0070529322
144Pr	1.0024881	-0.0053369322
145Pr	1.001207	-0.0066180322

15.60 Element 60 is Neodymium or Nd

Neodymium has 60 protons and 84 neutrons.

ANM = 0.9933403020

142Nd		0.0000000000
143Nd	1.002091	-0.0057340322
144Nd	1.000273	-0.0075520322
145Nd	1.0024863	-0.0053387322
146Nd	1.0005433	-0.0072817322
147Nd	1.0029835	-0.0048415322
148Nd	1.0007926	-0.0070324322
149Nd	1.003256	-0.0045690322
150Nd	1.000742	-0.0070830322
151Nd	1.002938	-0.0048870322
152Nd	1.000853	-0.0069720322

15.61 Element 61 is Promethium or Pm

Promethium has 61 protons and 84 neutrons.

ANM = 0.9932788337

143Pm		0.0000000000
144Pm	1.001658	-0.0061670322
145Pm	1.000158	-0.0076670322
146Pm	1.001947	-0.0058780322
147Pm	1.0004425	-0.0073825322

15.62 Element 62 is Samarium or Sm

Samarium has 62 protons and 88 neutrons.

ANM = 0.9935468580

140Sm		0.0000000000
141Sm	0.999481	-0.0083440322
142Sm	0.996722	-0.0111030322
143Sm	0.99943	-0.0083950322
144Sm	0.997371	-0.0104540322
145Sm	1.001411	-0.0064140322
146Sm	0.999631	-0.0081940322
147Sm	1.0018569	-0.0059681322
148Sm	0.9999248	-0.0079002322
149Sm	1.002362	-0.0054630322
150Sm	1.0000908	-0.0077342322
151Sm	1.0026569	-0.0051681322
152Sm	0.9998	-0.0080250322
153Sm	1.002365	-0.0054600322
154Sm	1.0001119	-0.0077131322
155Sm	1.0024309	-0.0053941322

15.63 Element 63 is Europium or Eu

Europium has 63 protons and 89 neutrons.

ANM = 0.9935816570

148Eu		0.0000000000
149Eu	0.999845	-0.0079800322
150Eu	1.001771	-0.0060540322
151Eu	1.0001482	-0.0076768322
152Eu	1.0018943	-0.0059307322
153Eu	0.9994858	-0.0083392322
154Eu	1.0017489	-0.0060761322
155Eu	0.9999141	-0.0079109322

15.64 Element 64 is Gadolinium or Gd

Gadolinium has 64 protons and 93 neutrons.

ANM = 0.9937973982

152Gd		0.0000000000
153Gd	1.0019585	-0.0058665322
154Gd	0.9991161	-0.0087089322
155Gd	1.0017564	-0.0060686322
156Gd	0.9995007	-0.0083243322
157Gd	1.0018374	-0.0059876322
158Gd	1.0001438	-0.0076812322
159Gd	1.0022848	-0.0055402322
160Gd	1.0006654	-0.0071596322
161Gd	1.0026151	-0.0052099322

15.65 Element 65 is Terbium or Tb

Terbium has 65 protons and 94 neutrons.

ANM = 0.9937948820

157Tb		0.0000000000
158Tb	1.0013885	-0.0064365322
159Tb	0.9999329	-0.0078921322
160Tb	1.0018216	-0.0060034322
161Tb	1.0004023	-0.0074227322

15.66 Element 66 is Dyprosium or Dy

Dyprosium has 66 protons and 97 neutrons.

ANM = 0.9939410214

154Dy		0.0000000000
155Dy	1.00133	-0.0064950322
156Dy	0.998529	-0.0092960322
157Dy	1.001183	-0.0066420322
158Dy	0.998943	-0.0088820322
159Dy	1.0013302	-0.0064948322
160Dy	0.9994583	-0.0083667322
161Dy	1.0017359	-0.0060891322
162Dy	0.999865	-0.0079600322
163Dy	1.0019328	-0.0058922322
164Dy	1.0004436	-0.0073814322
165Dy	1.0025285	-0.0052965322
166Dy	1.0011034	-0.0067216322

15.67 Element 67 is Holmium or Ho

Holmium has 67 protons and 98 neutrons.

ANM = 0.9939392341

163Ho		0.0000000000
164Ho	1.0014996	-0.0063254322
165Ho	1.0000886	-0.0077364322
166Ho	1.0019621	-0.0058629322
167Ho	1.0008488	-0.0069762322
168Ho	1.002387	-0.0054380322

15.68 Element 68 is Erbium or Er

Erbium has 68 protons and 99 neutrons.

ANM = 0.9939388486

158Er		0.0000000000
159Er	1.000791	-0.0070340322
160Er	0.998399	-0.0094260322
161Er	1.000912	-0.0069130322
162Er	0.998783	-0.0090420322
163Er	1.001255	-0.0065700322
164Er	0.999167	-0.0086580322
165Er	1.001526	-0.0062990322
166Er	0.9995671	-0.0082579322
167Er	1.0017551	-0.0060699322
168Er	1.000322	-0.0075030322
169Er	1.0022202	-0.0056048322
170Er	1.0008739	-0.0069511322
171Er	1.0025655	-0.0052595322

15.69 Element 69 is Thulmium or Tm

Thulmium has 69 protons and 100 neutrons.

ANM = 0.9939428608

169Tm		0.0000000000
170Tm	1.0015881	-0.0062369322
171Tm	1.000628	-0.0071970322

15.70 Element 70 is Ytterbium or Yb

Ytterbium has 70 protons and 103 neutrons.

ANM = 0.9940821218

164Yb		0.0000000000
165Yb	1.000791	-0.0070340322
166Yb	0.998602	-0.0092230322
167Yb	1.001068	-0.0067570322
168Yb	0.998947	-0.0088780322
169Yb	1.001293	-0.0065320322
170Yb	0.9995718	-0.0082532322
171Yb	1.001564	-0.0062610322
172Yb	1.0000557	-0.0077693322
173Yb	1.0018293	-0.0059957322
174Yb	1.0006513	-0.0071737322
175Yb	1.0024144	-0.0054106322
176Yb	1.0012952	-0.0065298322
177Yb	1.0026891	-0.0051359322

15.71 Element 71 is Lutetium or Lu

Lutetium has 71 protons and 104 neutrons.

ANM = 0.9940884088

170Lu		0.0000000000
171Lu	0.9994381	-0.0083869322
172Lu	1.0011729	-0.0066521322
173Lu	0.9998446	-0.0079804322
174Lu	1.0014069	-0.0064181322
175Lu	1.0004343	-0.0073907322
176Lu	1.0019145	-0.0059105322
177Lu	1.0010718	-0.0067532322

15.72 Element 72 is Hafnium or Hf

Hafnium has 72 protons and 106 neutrons.

ANM = 0.9941537404

174Hf		0.0000000000
175Hf	1.001463	-0.0063620322
176Hf	0.9998996	-0.0079254322
177Hf	1.0018121	-0.0060129322
178Hf	1.0004781	-0.0073469322
179Hf	1.0021173	-0.0057077322
180Hf	1.0007339	-0.0070911322
181Hf	1.0025512	-0.0052738322
182Hf	1.0014528	-0.0063722322

15.73 Element 73 is Tantalum or Ta

Tantalum has 73 protons and 108 neutrons.

ANM = 0.9942293375

180Ta		0.0000000000
181Ta	1.000531	-0.0072940322
182Ta	1.002156	-0.0056690322
183Ta	1.0012208	-0.0066042322
184Ta	1.0026354	-0.0051896322

15.74 Element 74 is Tungsten or W

Tungsten has 74 protons and 110 neutrons.

ANM = 0.9942898074

178W		0.0000000000
179W	1.001194	-0.0066310322
180W	0.999634	-0.0081910322
181W	1.001493	-0.0063320322
182W	1.0000072	-0.0078178322
183W	1.0020188	-0.0058062322
184W	1.0007082	-0.0071168322
185W	1.0024881	-0.0053369322
186W	1.0009448	-0.0068802322
187W	1.0027964	-0.0050286322
188W	1.0013285	-0.0064965322

15.75 Element 75 is Rhenium or Re

Rhenium has 75 protons and 111 neutrons.

ANM = 0.9943072854

185Re		0.0000000000
186Re	1.0020311	-0.0057939322
187Re	1.000767	-0.0070580322
188Re	1.0023613	-0.0054637322
189Re	1.0011146	-0.0067104322

15.76 Element 76 is Osmium or Os

Osmium has 76 protons and 114 neutrons.

ANM = 0.9944188118

184Os		0.0000000000
185Os	1.0015532	-0.0062718322
186Os	0.9997959	-0.0080291322
187Os	1.0019123	-0.0059127322
188Os	1.0000877	-0.0077373322
189Os	1.0023093	-0.0055157322
190Os	1.0002995	-0.0075255322
191Os	1.0024827	-0.0053423322
192Os	1.000551	-0.0072740322
193Os	1.0026709	-0.0051541322
194Os	1.0010305	-0.0067945322

15.77 Element 77 is Iridium or Ir

Ytterbium has 77 protons and 115 neutrons.

ANM = 0.9944354567

190Ir		0.0000000000
191Ir	1.000048	-0.0077770322
192Ir	1.002011	-0.0058140322
193Ir	1.0003214	-0.0075036322
194Ir	1.002152	-0.0056730322
195Ir	1.0009012	-0.0069238322
196Ir	1.0024204	-0.0054046322
197Ir	1.001253	-0.0065720322

15.78 Element 78 is Platinum or Pt

Platinum has 78 protons and 117 neutrons.

ANM = 0.9944823811

190Pt		0.0000000000
191Pt	1.001745	-0.0060800322
192Pt	0.999361	-0.0084640322
193Pt	1.0019494	-0.0058756322
194Pt	0.9996929	-0.0081321322
195Pt	1.0021108	-0.0057142322
196Pt	1.0001604	-0.0076646322
197Pt	1.0023887	-0.0054363322
198Pt	1.0005528	-0.0072722322
199Pt	1.0027	-0.0051250322

15.79 Element 79 is Gold or Au

Gold has 79 protons and 118 neutrons.

ANM = 0.9944778911

196Au		0.0000000000
197Au	0.9999987	-0.0078263322
198Au	1.0016736	-0.0061514322
199Au	1.0005229	-0.0073021322

15.80 Element 80 is Mercury or Hg

Mercury has 80 protons and 121 neutrons.

ANM = 0.9945809894

192Hg		0.0000000000
193Hg	1.001031	-0.0067940322
194Hg	0.998774	-0.0090510322
195Hg	1.001281	-0.0065440322
196Hg	0.999113	-0.0087120322
197Hg	1.00138	-0.0064450322
198Hg	0.999556	-0.0082690322
199Hg	1.0015109	-0.0063141322
200Hg	1.0000461	-0.0077789322
201Hg	1.0019763	-0.0058487322
202Hg	1.0003407	-0.0074843322
203Hg	1.0022295	-0.0055955322
204Hg	1.0006214	-0.0072036322

15.81 Element 81 is Thallium or Tl

Thallium has 81 protons and 123 neutrons.

ANM = 0.9946344381

200Tl		0.0000000000
201Tl	0.999856	-0.0079690322
202Tl	1.001287	-0.0065380322
203Tl	1.0002382	-0.0075868322
204Tl	1.0015193	-0.0063057322
205Tl	1.000564	-0.0072610322
206Tl	1.0016828	-0.0061422322
207Tl	1.0013087	-0.0065163322

15.82 Element 82 is Lead or Pb

Lead has 82 protons and 125 neutrons.

ANM = 0.9946739540

200Pb		0.0000000000
201Pb	1.001058	-0.0067670322
202Pb	0.999274	-0.0085510322
203Pb	1.001232	-0.0065930322
204Pb	0.9996526	-0.0081724322
205Pb	1.0014382	-0.0063868322
206Pb	0.9999835	-0.0078415322
207Pb	1.0014316	-0.0063934322
208Pb	1.0007552	-0.0070698322
209Pb	1.004438	-0.0033870322
210Pb	1.0030984	-0.0047266322
211Pb	1.0045485	-0.0032765322
212Pb	1.0031605	-0.0046645322
213Pb	1.0046835	-0.0031415322
214Pb	1.0032244	-0.0046006322

15.83 Element 83 is Bismuth or Bi

Bismuth has 83 protons and 126 neutrons.

ANM = 0.9946898494

207Bi		0
208Bi	1.0012715	-0.0065535322
209Bi	1.0006565	-0.0071685322
210Bi	1.0037217	-0.0041033322
211Bi	1.0031486	-0.0046764322
212Bi	1.0040167	-0.0038083322
213Bi	1.0030993	-0.0047257322
214Bi	1.004327	-0.0034980322
215Bi	1.003058	-0.0047670322
216Bi	1.004536	-0.0032890322

15.84 Element 84 is Polonium or Po

Polonium has 84 protons and 125 neutrons.

ANM = 0.9946010215

208Po		0
209Po	1.0011847	-0.0066403322
210Po	1.0004433	-0.0073817322
211Po	1.0037795	-0.0040455322

15.85 Element 85 is Astatine or At

Astatine has 85 protons and 125 neutrons.

ANM = 0.9945761621

210At		0
211At	1.0003483	-0.0074767322
212At	1.0032487	-0.0045763322

15.86 Element 86 is Radon or Rn

Radon has 86 protons and 136 neutrons.

ANM = 0.9951810656

Radon has no other isotopes with a half-life long enough for this analysis.

15.87 Element 87 is Francium or Fr

Francium has 87 protons and 136 neutrons.

ANM = 0.9951393978

222Fr		0.0000000000
223Fr	1.0021839	-0.0056411322

15.88 Element 88 is Radium or Ra

Radium has 88 protons and 138 neutrons.

ANM = 0.9951942534

226Ra		0.0000000000
227Ra	1.003768	-0.0040570322
228Ra	1.0018925	-0.0059325322

15.89 Element 89 is Actinium or Ac

Actinium has 89 protons and 138 neutrons.

ANM = 0.9951545234

225Ac		0.0000000000
226Ac	1.002868	-0.0049570322
227Ac	1.0016541	-0.0061709322
228Ac	1.003269	-0.0045560322

15.90 Element 90 is Thorium or Th

Thorium has 90 protons and 142 neutrons.

ANM = 0.9953084676

228Th		0.0000000000
229Th	1.0030209	-0.0048041322
230Th	1.0013718	-0.0064532322
231Th	1.0031705	-0.0046545322
232Th	1.001751	-0.0060740322
233Th	1.0035265	-0.0042985322
234Th	1.0020192	-0.0058058322

15.91 Element 91 is Protactinium or Pa

Protactinium has 91 protons and 140 neutrons.

ANM = 0.9951700433

Protactinium has no other isotopes with a half-life long enough for this analysis.

15.92 Element 92 is Uranium or U

Uranium has 92 protons and 146 neutrons.

ANM = 0.9954170221

232U		0.0000000000
233U	1.002479	-0.0053460322
234U	1.0013169	-0.0065081322
235U	1.0029778	-0.0048472322
236U	1.0016381	-0.0061869322
237U	1.0031622	-0.0046628322
238U	1.002058	-0.0057670322
239U	1.0035051	-0.0043199322

15.93 Element 93 is Neptunium or Np

Neptunium has 93 protons and 144 neutrons.

ANM = 0.9952808708

236Np		0.0000000000
237Np	1.0016034	-0.0062216322
238Np	1.002773	-0.0050520322
239Np	1.0019926	-0.0058324322

15.94 Element 94 is Plutonium or Pu

Plutonium has 94 protons and 150 neutrons.

ANM = 0.9955243398

236Pu		0.0000000000
237Pu	1.0023517	-0.0054733322
238Pu	1.0011502	-0.0066748322
239Pu	1.0026035	-0.0052215322
240Pu	1.0016501	-0.0061749322
241Pu	1.003038	-0.0047870322
242Pu	1.0018911	-0.0059339322
243Pu	1.0032604	-0.0045646322
244Pu	1.002201	-0.0056240322

15.95 Element 95 is Americium or Am

Americium has 95 protons and 148 neutrons.

ANM = 0.9953919124

241Am		0.0000000000
242Am	1.0027201	-0.0051049322
243Am	1.0018319	-0.0059931322

15.96 Element 96 is Curium or Cm

Curium has 96 protons and 151 neutrons.

ANM = 0.9954910656

243Cm		0.0000000000
244Cm	1.0013635	-0.0064615322
245Cm	1.0027386	-0.0050864322
246Cm	1.0017325	-0.0060925322
247Cm	1.0031303	-0.0046947322
248Cm	1.001995	-0.0058300322

15.97 Element 97 is Berkelium or Bk

Berkelium has 97 protons and 150 neutrons.

ANM = 0.9954085258

247Bk		0.0000000000
248Bk	1.002783	-0.0050420322

15.98 Element 98 is Californium or Cf

Californium has 98 protons and 153 neutrons.

ANM = 0.9955080643

249Cf		0.0000000000
250Cf	1.0015526	-0.0062724322
251Cf	1.0031809	-0.0046441322
252Cf	1.002039	-0.0057860322
253Cf	1.003507	-0.0043180322
254Cf	1.00219	-0.0056350322

15.99 Element 99 is Einsteinium or Es

Einsteinium has 99 protons and 153 neutrons.

ANM = 0.9954790968

252Es		0.0000000000
253Es	1.0018447	-0.0059803322
254Es	1.0031973	-0.0046277322
255Es	1.002251	-0.0055740322
256Es	1.003327	-0.0044980322
257Es	1.00238	-0.0054450322

15.100 Element 100 is Fermium or Fm

Fermium has 100 protons and 157 neutrons.

ANM = 0.9956216737

253Fm		0.0000000000
254Fm	1.0016694	-0.0061556322
255Fm	1.0031096	-0.0047154322
256Fm	1.00181	-0.0060150322
257Fm	1.003332	-0.0044930322

15.101 Element 101 is Mendelevium or Md

Mendelevium has 101 protons and 157 neutrons.

ANM = 0.9955930111

258Md		0.0000000000
259Md	1.002079	-0.0057460322
260Md	1.00314	-0.0046850322

15.102 Element 102 is Nobelium or No

Nobelium has 102 protons and 157 neutrons.

ANM = 0.9955597243

253No		0.0000000000
254No	1.000392	-0.0074330322
255No	1.002235	-0.0055900322
256No	1.001092	-0.0067330322
257No	1.002605	-0.0052200322
258No	1.001322	-0.0065030322
259No	1.00282	-0.0050050322

15.103 Element 103 is Lawrencium or Lr

Lawrencium has 103 protons and 159 neutrons.

ANM = 0.9956203250

260Lr		0.0000000000
261Lr	1.00137	-0.0064550322
262Lr	1.00273	-0.0050950322

15.104 Element 104 is Rutherfordium or Rf

Rutherfordium has 104 protons and 157 neutrons.

ANM = 0.9955093417

261Rf		0
262Rf	1.00116	-0.0066650322
263Rf	1.00257	-0.0052550322

15.105 Element 105 is Dubnium or Db

Dubnium has 105 protons and 157 neutrons.

ANM = 0.9954932587

262Db		0.0000000000
263Db	1.00092	-0.0069050322
266Db		
267Db	1.00144	-0.0063850322
268Db	1.0032	-0.0046250322

The Dubnium isotopes missing from the list have a half-life too brief for a valid measurement.

15.106 Element 106 is Seaborgium or Sg

Seaborgium has 106 protons and 157 neutrons.

ANM = 0.9955782911

266Sg		0.0000000000
267Sg	1.00238	-0.0054450322

15.107 Element 107 is Borhrium or Bh

Borhrium has 107 protons and 157 neutrons.

ANM = 0.9954605831

Borhrium has no other isotopes with a half-life long enough for this analysis.

15.108 Element 108 is Hassium or Hs

Hassium has 108 protons and 159 neutrons.

ANM = 0.9955129970

Hassium has no other isotopes with a half-life long enough for this analysis.

15.109 Element 109 is Meiterium or Mt

Meiterium has 109 protons and 159 neutrons.

ANM = 0.9955076823

Meiterium has no other isotopes with a half-life long enough for this analysis.

15.110 Element 110 is Damstadtium or Ds

Damstadtium has 110 protons and 152 neutrons.

ANM = 0.9959284003

Damstadtium has no other isotopes with a half-life long enough for this analysis.

15.111 Element 111 is Roehtgenium Rg

Roehtgenium has 111 protons and 161 neutrons.

ANM = 0.9959095990

Roehtgenium has no other isotopes with a half-life long enough for this analysis.

15.112 Element 112 is Copernicium or Cn

Copernicium has 112 protons and 173 neutrons.

ANM = 0.9959578982

Copernicium has no other isotopes with a half-life long enough for this analysis.

15.113 Element 113 is Nihonium or Nh

Nihonium has 113 protons and 173 neutrons.

ANM = 0.9959420888

Nihonium has no other isotopes with a half-life long enough for this analysis.

15.114 Element 114 is Flerovium or Fl

Flerovium has 114 protons and 175 neutrons.

ANM = 0.9903316268

Flerovium has no other isotopes with a half-life long enough for this analysis.

15.115 Element 115 is Muscovium or Mc

Muscovium has 115 protons and 173 neutrons.

ANM = 1.0074919150

Muscovium has no other isotopes with a half-life long enough for this analysis.

15.116 Element 116 is Livermorium or Lv

Livermorium has 116 protons and 177 neutrons.

ANM = 1.0249813155

Livermorium has no other isotopes with a half-life long enough for this analysis.

15.117 Element 117 is Tennessine or Ts

Tennessine has 117 protons and 177 neutrons.

ANM = 0.9960165606

293Ts		0.0000000000
294Ts	1.00222	-0.0056050322

15.118 Element 118 is Oganesson or Og

Oganesson has 118 protons and 176 neutrons.

ANM = 0.9960050068

Oganesson has no other isotopes with a half-life long enough for this analysis.

16 Chart Isotope Changes

The behavior of the change in each isotope's mass when adding 1 neutron is charted below. The change in mass is compared to the sum of a proton and electron. That sum is the expected mass of a neutron or Mn

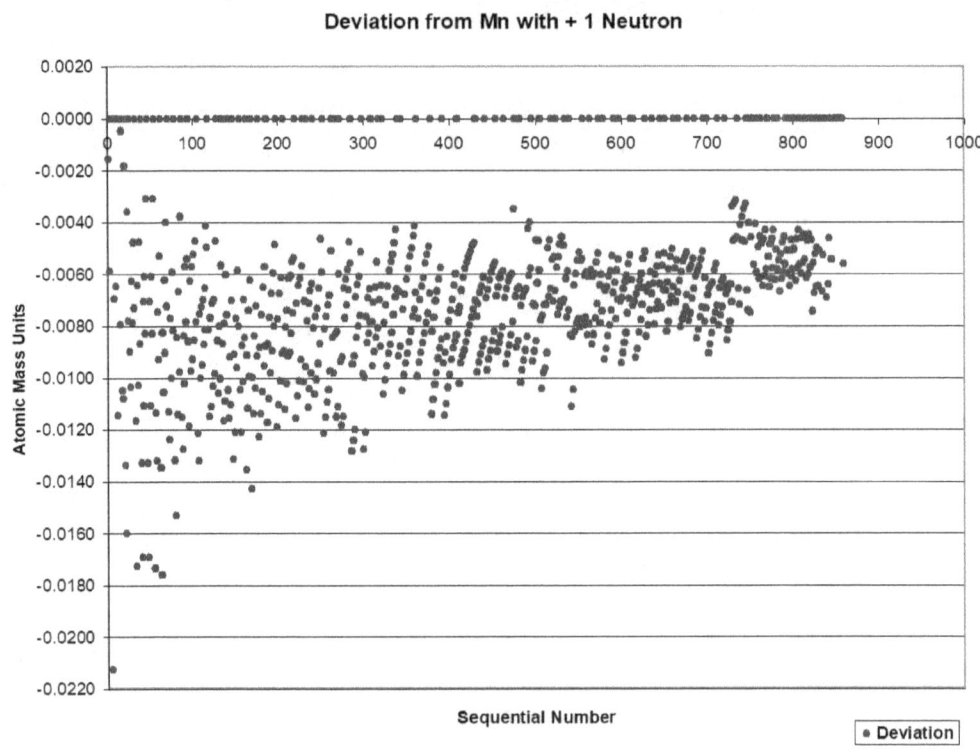

In a single image the minor change in the mass of each neutron is clearly visible.

That the atomic mass defect has remained unexplained suggests no one ever plotted this behavior like in this image.

The change in atomic mass when adding a single neutron should be consistent for all isotopes according to the Standard Model.

Instead, each element has slightly different deviation but nearly all are within a range of values, with most lying between about -0.0140 and -0.0030

To put this deviation in its context, the sum of an electron and proton, or a neutron, is

1.0078250322 amu

The marks in the chart are the tiny deviations from that value.

The first isotope of each element has a deviation of 0.00 since it has no preceding isotope to compare. The chart has many marks on 0.00 but none of them are from an actual measured deviation. There is no isotope having a change in mass from the preceding isotope exactly the value shown above. All of the isotope values are non-zero.

The list of isotopes for creating this chart was by atomic number, then by its isotopes.

That outlier at the bottom left is the change from ^{3}He to ^{4}He so the mass change was rather different when adding the second neutron to 2 protons. The chart is only an overview. All the details are with each element in section 15 Periodic Table.

17 Elements Conclusion

Mass defect presents a challenge because it originates in the atomic nucleus.

Each atom has a history of its formation. That involved the 2 fundamental particles of proton and electron, but large nuclei probably included fusing smaller nuclei to create a further combination. The current solar model based on only fusion in the core cannot explain the observed distribution of many elements. As noted in the author's book Cosmology Transition, a new solar model is being recognized which can invoke other mechanism for creating elements. A solar model is out of scope for this book but one must note the creation of elements is more complicated than the current fusion model implies. That model is known to be unable to explain the distribution of some elements. Its assumed fusion sequence inevitably encounters unstable isotopes among the steps. The creation sequence of the observed nuclei is out of scope for this book, but that uncertainty is relevant. Currently, some students are taught the mass defect leads to a mass to energy calculation which requires precise measurements for the correct change in mass. This lesson never teaches the mechanism behind the change in mass, and probably misses the importance of consistent precision of the values.

The analysis of mass defect using many isotopes results in a change in our understanding of protons and neutrons.

The proton is a fundamental particle. The discovery that it can be broken into fragments contributed nothing to our understanding of atoms.

The neutron is simply a proton with an adjacent electron.

Physicists claim to have observed 3 fragments from breaking a neutron. The most likely explanation is these are the same 3 fragments as claimed from a proton. These fragments have no defined attributes. Therefore, the situation is this:

Proton breaks into 3 fragments which do almost nothing. Neutron breaks into 3 fragments which do almost nothing.

The quarks are claimed to account for only 1% of the proton's mass so they cannot explained a proton's observed behaviors.

There is no evidence suggesting these two sets of 3 fragments are not the same sets.

Physicists must provide evidence the 2 sets are actually different. Until then, one must assume they are just debris with their functions broken and disabled.

Because a neutron is neutral, it cannot be manipulated by electric and magnetic fields in an accelerator. A neutron loses its electron in a few minutes when outside an atom's nucleus.

Some very heavy elements eject neutrons during their radioactive decay sequence. This source is often used for neutron experiments.

Whenever a proton or neutron is forced by compression into a nucleus, the size of the proton is slightly reduced. This causes a corresponding slight reduction in its reactivity to other masses.

This is observed a reduction in the proton's measured mass. This is a behavior of a coherent particle where it appears affected by its change in volume.

This is definitely not a behavior expected from 3 independent particles.

Excerpt from Wikipedia:

In quantum chromodynamics, the modern theory of the nuclear force, most of the mass of protons and neutrons is explained by special relativity. The mass of a proton is about 80–100 times greater than the sum of the rest masses of the quarks that make it up, while the gluons have zero rest mass. The extra energy of the quarks and gluons in a region within a proton, as compared to the rest energy of the quarks alone in the QCD vacuum, accounts for almost 99% of the mass. The rest mass of a proton is, thus, the invariant mass of the system of moving quarks and gluons that make up the particle, and, in such systems, even the energy of massless particles is still measured as part of the rest mass of the system.

The internal dynamics of protons are complicated, because they are determined by the quarks' exchanging gluons, and interacting with various vacuum condensates. Lattice QCD provides a way of calculating the mass of a proton directly from the theory to any accuracy, in principle. The most recent calculations claim that the mass is determined to better than 4% accuracy, even to 1% accuracy. These claims are still controversial, because the calculations cannot yet be done with quarks as light as they are in the real world.

This means that the predictions are found by a process of extrapolation, which can introduce systematic errors. It is hard to tell whether these errors are controlled properly, because the quantities that are compared to experiment are the masses of the hadrons, which are known in advance.

(Excerpt end)

Observation:

Even after so many years since quarks were found as debris from particle collisions, a proton's mass using quarks remains a problem, with accuracy in the range of 1 to 4%, when the result is "known in advance."

There are 2 particles in a nucleus, the proton and neutron. This book uses data which enable only the neutron behavior to be measured.

Since a neutron is a proton having an attached electron, this description refers to only a proton for simplicity.

Observational data confirm both the size reduction and the mass reduction when these 2 particles are in a nucleus.

With a proton as a single entity having no pieces, it is reasonable to expect its reactivity to other masses could be reduced as its volume is reduced. With the proton having 3 individual pieces interacting with "vacuum condensates" for its mass behavior, it is reasonable to expect a reduction in the proton's volume should not affect the 3 quarks in how they perform their 2 distinct tasks of mass and charge. The observation a proton's volume affects its mass cannot be explained by the Standard Model using quarks. One could suspect the physicists never noticed this change was being measured within the nucleus. The Standard Model cannot explain how its set of 3 quarks can change the proton's mass as observed.

The Standard Model simply fails to explain a crucial behavior in every element. It is not logical to expect 3 disjointed pieces can execute 2 separate unrelated tasks. The claim a proton has 3 quarks having no measurable features must be dropped. The Standard Model should be fixed by dropping all its false quasi-particles like quarks and taking a practical approach driven by the accumulated data.

There is fundamental uncertainty when dealing with mass defect. This slight reduction in a proton's apparent mass occurs during the process of fusing particles together.

No atom has spontaneously appeared. Intergalactic Birkelund currents bring protons and electrons, as plasma filaments, to galaxies in the universe's electrical network.

Stars are composed of liquid metallic hydrogen, which is a name for a lattice of protons being maintained by free electrons. Stars have the great electrical energy at their surface, the photosphere, to compress particles together in a process called transmutation, or a Low Energy Nuclear Reaction, which is sometimes called cold fusion, as temperature is not critical in the process.

All the observed atoms were assembled, over a time, by fusing combinations of protons and electrons.

18 Final Conclusion

Practical Particle Physics proposes the Standard Model be updated, by removing its quarks, its quasi-particles and its use of relativity. The result is a simpler atomic model with a very small set of fundamental subatomic particles, the proton, electron, neutrino, and (maybe) muon. Anti-particles are not fundamental particles.

All atomic behaviors currently assumed to be photon particle-like behaviors can be explained as light wave length behaviors.

Quarks have yet to provide a useful function, beyond a reason for using the LHC. LHC is not the correct justification to keep the quarks in the model. Quarks are in the class of objects found in the debris of particle accelerators and are not fundamental particles.

Newton's force of gravity and Maxwell's electromagnetic forces also return to physics, after dropping such misdirected concepts like photon and graviton. Neither quasi-particle exists. The 3 fundamental forces of electric, magnetic, and gravitational are well tested and confirmed over many years. Relativity was a mistake and a wasteful diversion for several sciences. Relativity was explicitly for only a special moving observer, and nothing in the universe has such a special observer commanding its motion using its reference frame. Everything in the universe moves by the combination of forces moving on it. Nothing moves by commanded motion within its reference frame.

After an update to its atomic model, particle physics has a practical basis on the substantial data accumulated by measurements, often taken during experiments which enable conflicting data to fix wrong assumptions.

Theories must be based on the evidence.

Science is not the practice of forming a theory, then seeking evidence which conforms. That technique misses conflicting evidence.

When conflicting evidence arises later, the theory must be changed to suit the new set of evidence, but this correction is not always performed.

Particle physics did not correct its theories based on subsequent precise measurements of atomic behaviors.

This book has recommended changes to the theories of particle physics to suit the evidence.

19 References

The references in the book are available as clickable links from a page in the author's web site.

1. Start web browser
2. Go to this site: www.cosmologyview.com
3. Make sure the browser is on the correct home page:

Cosmology Views

4. Scroll to near the middle.
5. Select: **Books by the author**

This page presents information for each book.

Locate the columns for this book.

6. Locate: Practical Particle Physics
7. Below it, locate the date of this book's edition:

12/26/2020 References

8. Select: **References** after the correct date.

The selected page will list the references in the book by page number, with a link to that reference.

Each link has the page number and the link's target indicating whether it is to a pdf, a zip (only for the spreadsheet with element data), a YouTube video, or a URL link to a web page. The user is aware of what the browser will do with the link.

www.ingramcontent.com/pod-product-compliance
Lightning Source LLC
Chambersburg PA
CBHW070616220526
45466CB00001B/23